THE INVISION GUIDE TO

SEXUAL HEALTH

ALSO BY ALEXANDER TSIARAS

THE INVISION GUIDE TO A HEALTHY HEART

FROM CONCEPTION TO BIRTH: A LIFE UNFOLDS

THE ARCHITECTURE AND DESIGN
OF MAN AND WOMAN:
THE MARVEL OF THE HUMAN BODY,
REVEALED

THE INVISION GUIDE TO

SEXUAL HEALTH

ALEXANDER TSIARAS

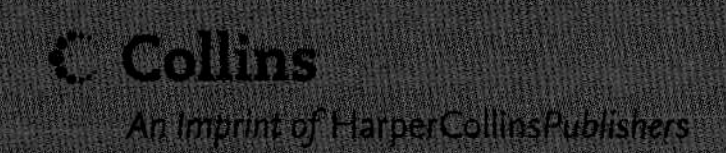
Collins
An Imprint of HarperCollinsPublishers

FIRST EDITION

Designed by Cindy Goldstein, Eric Baker Design Associates

Library of Congress Cataloging-in-Publication Data
has been applied for.

ISBN-10 0-06-087909-2
ISBN-13 978-0-06-087909-9

06 07 08 09 10/QW 10 9 8 7 6 5 4 3 2 1

This book contains advice and information relating to health care. It is not intended to replace medical advice and should be used to supplement rather than replace regular care by your doctor. It is recommended that you seek your physician's advice before embarking on any medical program or treatment.

All efforts have been made to ensure the accuracy of the information contained in this book as of the date published. The authors and the publisher expressly disclaim responsibility for any adverse effects arising from the use or application of the information contained herein.

ACKNOWLEDGMENTS

THIS BOOK IS DEDICATED WITH GRATITUDE TO:

Attila Ambrus, Creative Director/Director of Scientific Visualization/Anatomical Travelogue Inc.:
For his ability to multi-task, multi-task, and multi-task, and for always delivering quality.

Matt Wimsatt, Project Manager/Visualization Expert:
For never missing a detail, for his combined creative and organizational skills and leadership, and for his ability to (in the late evening and weekends) galvanize his team.

Jeremy Mack, Creative Director,
Department of Scientific Visualization:
For his continued creativity, advancing technical expertise, artistic supervision, and uplifting nature.

Medical visualization specialists:
Laszlo Balogh, Ann Canapary, Laura Gibson, Ezra Kortz, Mark Mallari, Taryn McLaughlin, Karina Metcalf, Jean-Claude Michel, Gloria Situ, Jacquelyn Sun, Casey Steffen, and Poy Yee, for their artistry, attention to detail, long hours, and individual and collective creativity.

Designers:
Senior Designer Poy Yee, and Associate Designer Jacquelyn Sun, for their creativity and design expertise.
Eric Baker Design, with special thanks to Eric Baker and Cindy Goldstein for their continued originality and ability to work under demanding deadlines.

Medical visualization intern: Shlomo Sam Spaeth.

To the participating AT staff:
Nilufer Candar, Sharon Ching, Stewart Deitch, and Ildiko McGivney

Models:
Attila Ambrus, Andres Arango, Sharon Ching, Gwen DeCelles, Megan Metcalf, Timea Resan, Gavin Whelan

Researchers, writers and proofreaders:
Sharon Ching, Alison Dalton, Shirley Chan

Toni Sciarra at Collins:
For her patience and her thorough professionalism.

National Institutes of Health
National Library of Medicine and the Visible Human project

National Museum of Health and Medicine of the AFIP
Adrianne Noe, Director of the National Museum of Health and Medicine of the AFIP

Levine Plotkin & Menin, LLP and Bob Levine for his belief in our work and vision

Scientific Visualization Software developed in collaboration with Volume Graphics, GmbH, Germany
(www.volumegraphics.com)

Stock imagery is provided by the Science Source division of Photo Researchers Photo Researchers Inc.
(www.sciencesource.com)
Graham McKenzie-Smith, photographer of "abstinence"–image in Chapter 6

Elizabeth Boskey, Ph.D., for her ability to write so eloquently and research across so many disciplines.

E. Scott Pretorius, M.D.
University of Pennsylvania, Department of Radiology, Wallace T. Miller, Sr. Chair of Radiologic Education and Associate Chairman, Hospital of the University of Pennsylvania
Lucy Brown, Ph.D.
Departments of Neurology and Neuroscience, Albert Einstein Medical Center

To the scientific advisory board:
Helen Fisher, Ph.D., Department of Anthropology, Rutgers University

Mark J. Holterman M.D., Ph.D., F.A.A.P., F.A.C.S., Associate Professor of Surgery and Division Chief, Pediatric Surgery, University of Illinois College of Medicine at Chicago

Barry R. Komisaruk, Ph.D., Professor of Psychology, Rutgers University, Former MBRS Program Director, National Institutes of Health

Judith H. LaRosa, Ph.D., R.N., Professor of Preventive Medicine and Community Health, Deputy Director, Master of Public Health Program, SUNY Downstate Medical Center

Dawn P. Misra, M.H.S., Ph.D., Associate Professor, Health Behavior & Health Education, Director, Reproductive and Women's Health Interdepartmental Concentration, University of Michigan School of Public Health

Carlos Beyer, Ph.D., Professor and Director, Center for Investigation of Animal Reproduction (CIRA) University of Tlaxcala, Tlacala, Mexico

Beverly Whipple, PhD, RN, FAAN Professor Emerita, Rutgers UniversitySecretary General, World Association for Sexual Health (WAS) (2005-2009) Past-president, Society for the Scientific Study of Sexuality (SSSS) Past-president, American Association of Sex Educators, Counselors andTherapists (AASECT)Past-Director, International Society for the Study of Women's Sexual Health (ISSWSH)

These individuals were instrumental in the accuracy and direction of the science and imagery that have pushed the limits of Anatomical Travelogue's technology. We thank them.

PROLOGUE

Talking about sex isn't easy. Over the past several years, I've had conversations with many teachers and parents about how they communicate about sex with their students and children. Universally, they have expressed frustration in initiating conversations and disappointment in the lack of compelling educational material that will both engage and accurately inform. Their frustration is multiplied by the knowledge that intelligent communication about sex is increasingly vital in the age of HIV/AIDS.

The InVision Guide to Sexual Health is an answer to these frustrations. . . and it goes further. This book will help to stimulate conversations between parent and child or teacher and student, but it also provides engaging knowledge for lovers and solutions to problems that are often addressed individually by the many disciplines that treat and study sex. Our multidisciplinary approach unites the knowledge of urologists, endocrinologists, neurobiologists, sociologists, and anthropologists—all in one book.

We hope *The InVision Guide to Sexual Health* will answer questions and provide illumination about many of the important issues of sex: the sexual changes of the body throughout a lifetime, sexual stimulation, sexually transmitted diseases, and sexual disorders such as erectile dysfunction.

This book is uniquely engaging in its visual representations, and it is also rich in information. Our hope is that we may help to produce open and responsible conversations about the beauty and intricacies of the biology of sex.

—ALEXANDER TSIARAS

FOREWORD

Sex. No other basic human urge inspires such deep emotion, drives us with such force, or inspires so much art and literature. Often regarded as taboo, cloaked in secrecy, and riddled with misconceptions, sex—even with dramatic advances in scientific knowledge and techniques of investigation—still remains a mystery.

How can such a powerful force be comprehended? How can we understand sex in a manner that conveys its richness and depth, its drama and its beauty? Alexander Tsiaras and his talented group at Anatomical Travelogue have found a way. With sophisticated manipulation of complex medical data, Tsiaras and his team create breathtaking three-dimensional images of the human body. Slices derived from MRIs, CTs and 3D-microscopy of actual human bodies are enhanced digitally to create visuals that can be understood immediately—information that otherwise would require volumes of text to convey. The rich, fluid, dynamic images truly allow us to see and understand sex—literally inside and out.

The textual information contained in this book is thoroughly researched and is based in scientific fact. Readers will learn about their bodies and sexual development, their hormones and erotic attraction, and sexual diseases and disorders. The immediately accessible visual narratives personalize human sexuality—giving intimacy to the information and making sexual health and well-being natural and essential.

The compelling images and text in this book can also serve as catalysts for conversations about sex—parent to child, teacher to student, adult to adult—guiding discussions that might otherwise be awkward. And certainly this stunning book humanizes sexual information, vividly showing the elegance, intricacy and vulnerability of the human body.

As the powerful images and comprehensive text dispel misconceptions and illuminate the physical landscape of human desire and reproduction, one awakens to even greater wonders: the miracle of the human body—and the eternal dance of sex.

Helen Fisher, Ph.D.,
Department of Anthropology, Rutgers University

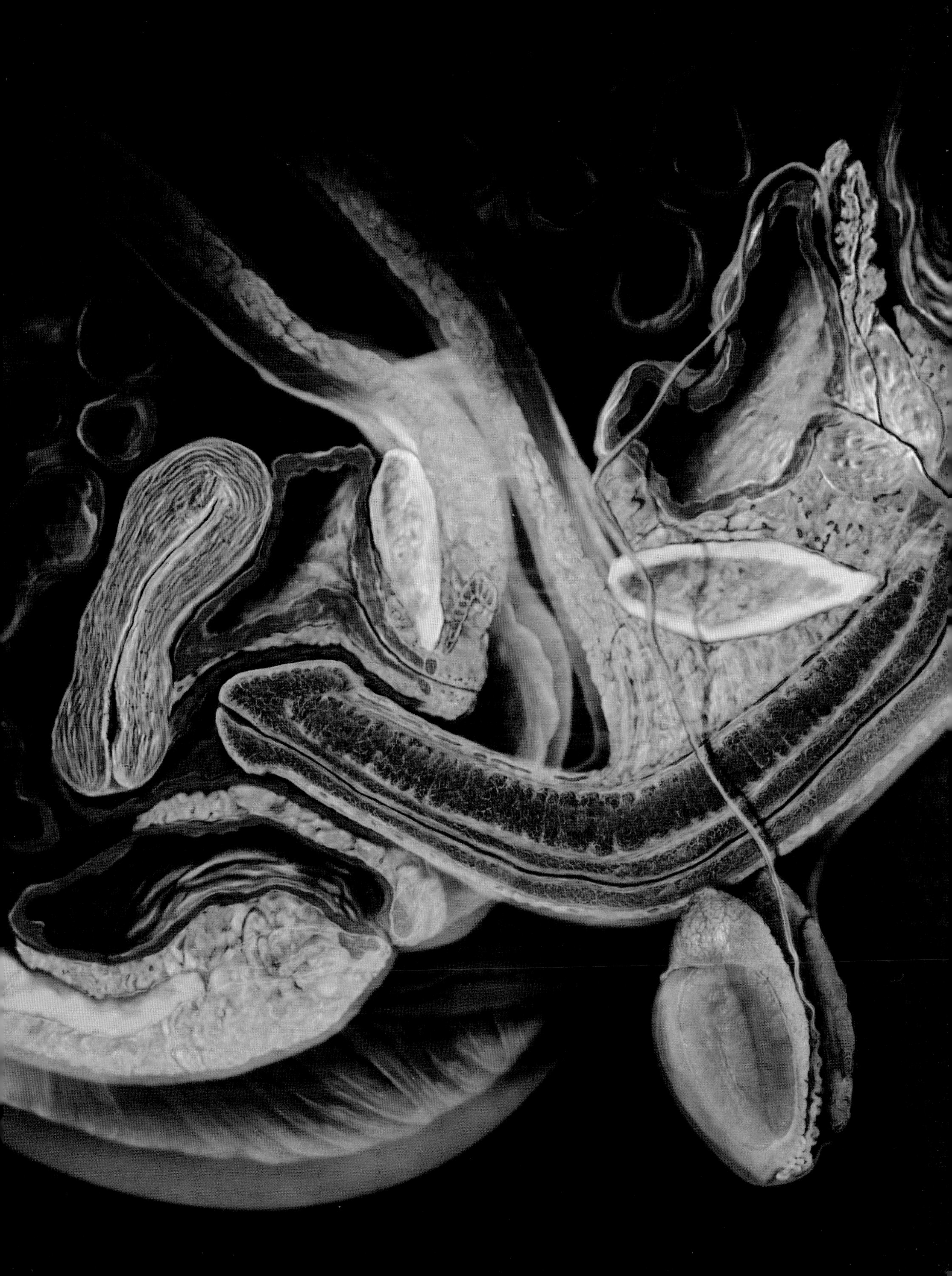

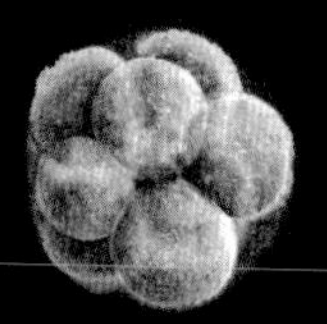

THE BODY DOES NOT INCREASE GREATLY IN SIZE DURING THE EMBRYONIC PERIOD. BUT DURING THE THE FETAL PERIOD, BEGINNING AT WEEK NINE, A PHASE OF RAPID GROWTH STARTS THAT CONTINUES UNTIL AFTER BIRTH.

1

THE MARVEL OF SEXUAL DEVELOPMENT

Let's talk about sex. Why should something that is as integral a part of life as breathing be so difficult to discuss openly? We think about sex from the time our infant hands first discover that our genitals are as fun to play with as our toes (and are easier to reach!), yet it's a subject that is shrouded in mystery to most children and even many adults. But not in these pages. . . . Read on, and open your eyes to the miracles of the human body at its most vulnerable—and its most powerful.

Sex is our deepest form of consciousness.
—D. H. LAWRENCE, *Fantasia of the Unconscious*

LET'S HAVE A CONVERSATION

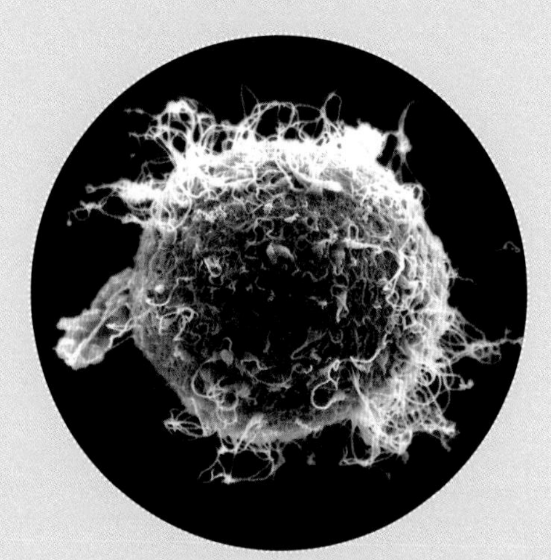

BREAK EGG, ADD SPERM: What is the "recipe" for a human being? Mom may want a daughter, but it's up to Dad whether she has one or a son.

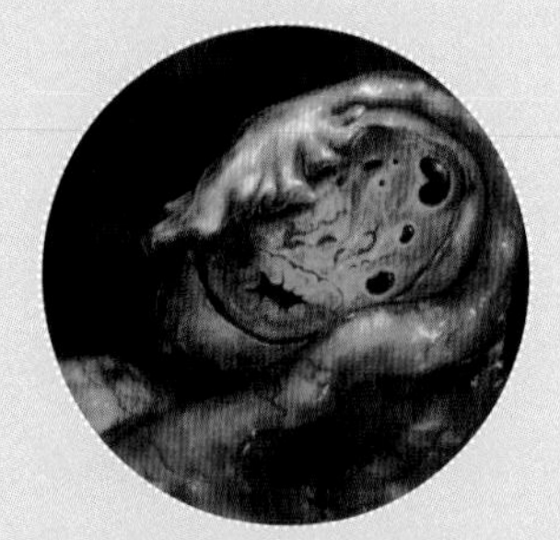

PINK BLANKET OR BLUE?: Every human being starts out as a female. But almost half the time, a biological switch is thrown and a male is born. How does this happen, and are men and women really that different?

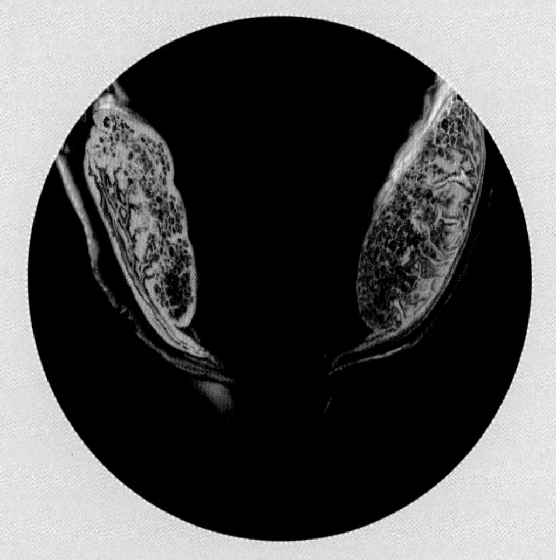

THE SAME, BUT DIFFERENT: Little boys and little girls may look very different, but just like legends of true love, every part you find on one has a perfect match upon the other. *Homology* is just a two-dollar word to describe how very much alike we are, both above and underneath the skin.

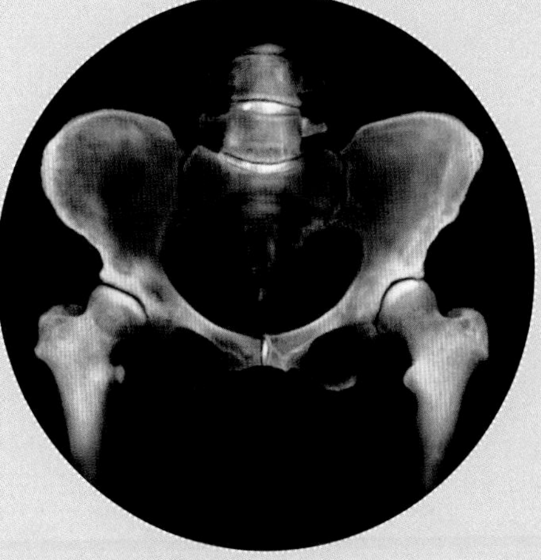

THAT TIME OF THE MONTH: Nothing marks puberty so clearly for a girl as the moment she looks down to discover blood staining her clothing—at home or in the middle of class. Why is that monthly visitor so important?

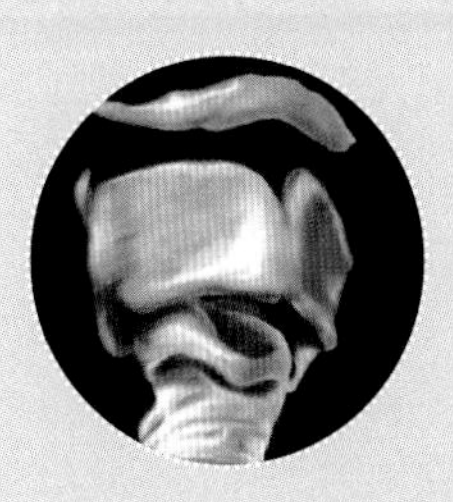

CH-CH-CH-CHANGES: From boys to men, one of the most striking changes that occurs during male puberty can be the way a young man sounds. It's not only the external organs that grow when the hormonal flood begins.

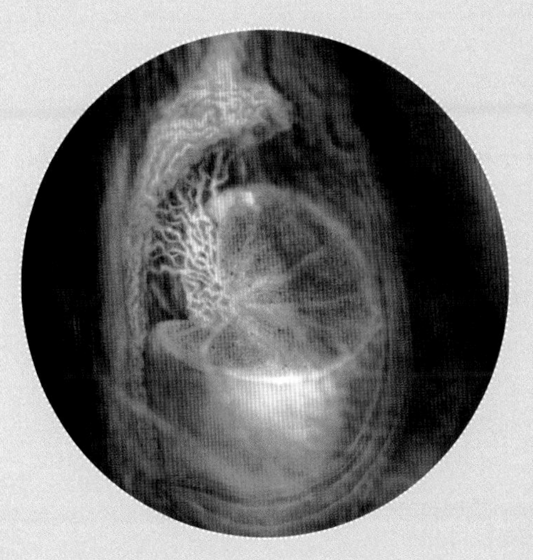

A RISING CONCERN: A young man's first public erection is a difficult, if exciting, time in his life. What quirk of biology suddenly makes every thought a catalyst for sexual arousal?

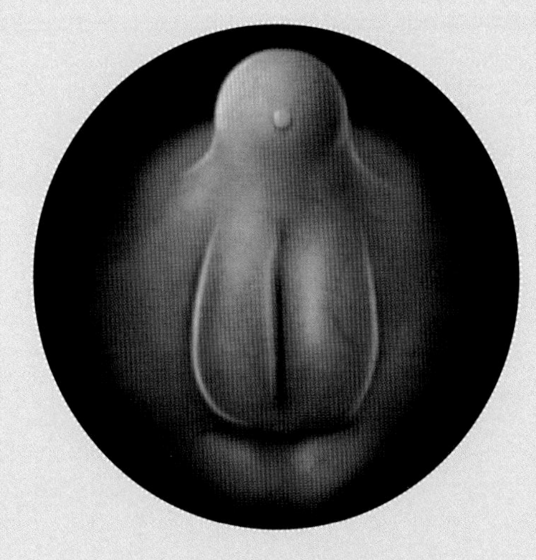

IT'S A (FE)MALE THING: Hormones are not just a "female thing." They create the sexual and emotional environment in men as well, and they have profound effects on life not just during "that time of the month," but in every moment from conception to death.

6.
XX
XY

Sex education is legitimate
in that girls cannot be taught soon
enough how children
don't come into the world.
—KARL KRAUS,
Half-Truths and One-and-a-Half Truths

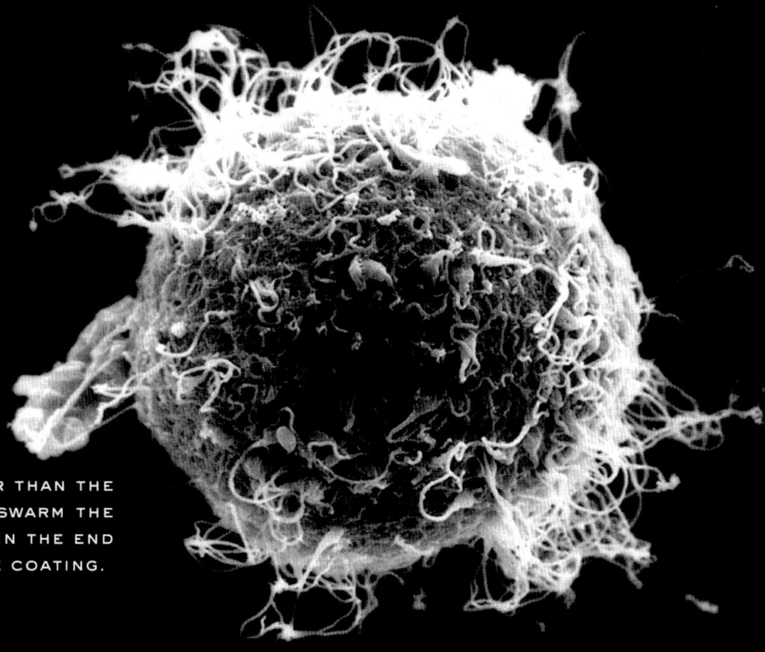

THE HUMAN EGG IS 100,000 TIMES LARGER THAN THE HUMAN SPERM. MILLIONS OF SPERM MAY SWARM THE EGG IN AN ATTEMPT TO FERTILIZE IT, BUT IN THE END ONLY ONE MAY PENETRATE ITS PROTECTIVE COATING.

RECIPE FOR A BABY

Approximately once a month, a woman's body gears up for reproduction. Her ovary releases an egg. While her body waits to see if that egg is destined to become an embryo, her uterus prepares so that it will be ready to incubate the potential life from conception to birth. The egg can be fertilized only during the brief window of time in which it is viable. If a sperm wriggles through the egg's sugary shell during this time, an embryo is formed. That embryo, given a bit of luck and nine months of living the high life in its mother's womb, will be born into the world as a son . . . or daughter.

One of the many seemingly magical things about human reproduction is that only one sperm can ever penetrate an egg. Many will attach to it, striving not to have their genetic material wasted after their exhausting journey through the female reproductive tract. There they swim valiantly and frantically in a hostile environment that only the strongest sperm can survive. But only one will eventually get through. As soon as it does, the egg changes, sealing off its borders in a process known as the zonal reaction, leaving the remaining sperm to their doom.

This is the first sign of how much the woman controls in human reproduction, but there is one aspect that is biologically forever out of reach: her body has no say in the sex of her future child. All eggs contain a single X chromosome, common to both men and women. It's the sperm that contain the complementary X or Y that will determine whether the child is a boy (XY) or a girl (XX). The man's genes determine a child's sex.

Sex, however, is not just for reproduction—not in humans. That may be its primary biological mandate, but it serves other purposes as well; otherwise it wouldn't be so much fun. It helps strengthen relationships, encourages a man to stick around to care for his child, and affects not only health but also well-being. It makes us fascinating to others, and we continue to explore its fascination to us.

EVERYONE STARTS OUT FEMALE

Unisex isn't just a term for clothing; it's also an accurate description of human beings during their earliest development in the womb. For more than a month, all embryos, whether they are destined to be male or female, grow in a way that looks exactly the same. They take after their mother. It's only natural, really, since she provides the vast majority of the support structure in which they grow. The cytoplasm of the egg, a complex bath of essential nutrients; mitochondria, the biological engines that provide the energy of life; and the womb, the incubator in which the embryo starts to thrive—all are given by Mom. At this stage of development, the father has provided his DNA, a smidgen of cytoplasm, a structure called the centrosome—which is necessary for the fertilized egg to undergo cell division—and, at the site of sperm penetration, the point of reference around which the left and right sides of the body will eventually form.

Adam may have been created first in the Bible, but it's Eve who is the original child of every womb. All embryos, male (XY) or female (XX), start off along the developmental pathway to womanhood, and for the first six weeks of growth they look exactly the same. In order for a boy to be formed, the Y chromosome must not only send out the correct chemical signals to the brain and body, but those signals must be properly received.

The definition of biological sex is a difficult one. You might think that a look at the 23rd chromosome pair should be enough to tell us whether a child is a boy or a girl, but sometimes it's not that simple. When we look at a newborn baby, we don't take blood and examine the shapes of the chromosomes to figure out if it's male or female; we look at the child and know. Or at least we think we do. Most of the time, what you see is what you get, but the process of sexual development is complex and can occasionally lead to subtle or fascinating surprises. If some sex-determining genes did not function correctly, or if the growing fetus is exposed to an external source of hormones, what you see *isn't* always what you get. A child who, at birth, appears to be a boy may actually have the genes of a girl. The girl you expect to see grow into a woman may turn out, at puberty, to actually be a man. Life is full of surprises, and one of them is how little difference there is between the sexes. After all, even by the most generous estimate, the number of genes unique to men represents only around one percent of the human genetic code . . . that's far less than the relative size of an extra rib.

THE HUMAN EMBRYO AT FIVE WEEKS OF DEVELOPMENT. AT THIS POINT IT IS IMPOSSIBLE TO VISIBLY DISTINGUISH BETWEEN MALE AND FEMALE, ALTHOUGH THE GONADAL RIDGE, WHICH WILL BECOME THE GENITALIA, IS VISIBLE.

The spark that starts male development along its unique pathway comes from the sex-determining region Y (SRY) gene on the Y chromosome. This gene's products, unique to male fetuses, kick off the development of the testicles at around the 6th week of development. If SRY is not present, or not working, by its second month the fetus will start to grow a uterus and continue along the road to womanhood.

By the 8th week of development, the fetal gonads—the organs that will develop into the ovaries or testicles—can be clearly determined. The development of the gonads is critical to the correct formation of the male and female sex organs. Before this point, the internal genitalia consisted of two undifferentiated sets of ducts: the Mullerian ducts and the Wolffian ducts. In healthy fetuses, the Mullerian ducts will grow into the uterus, fallopian tubes, and inner vagina—unless anti-mullerian hormone (AMH) is produced by the testicles to tell them to degenerate. The name AMH clearly describes its purpose: it tells the Mullerian ducts to wither. Without it, since the Mullerian ducts need no hormonal stimulation

THE UPS AND DOWNS OF BEING MALE OR FEMALE

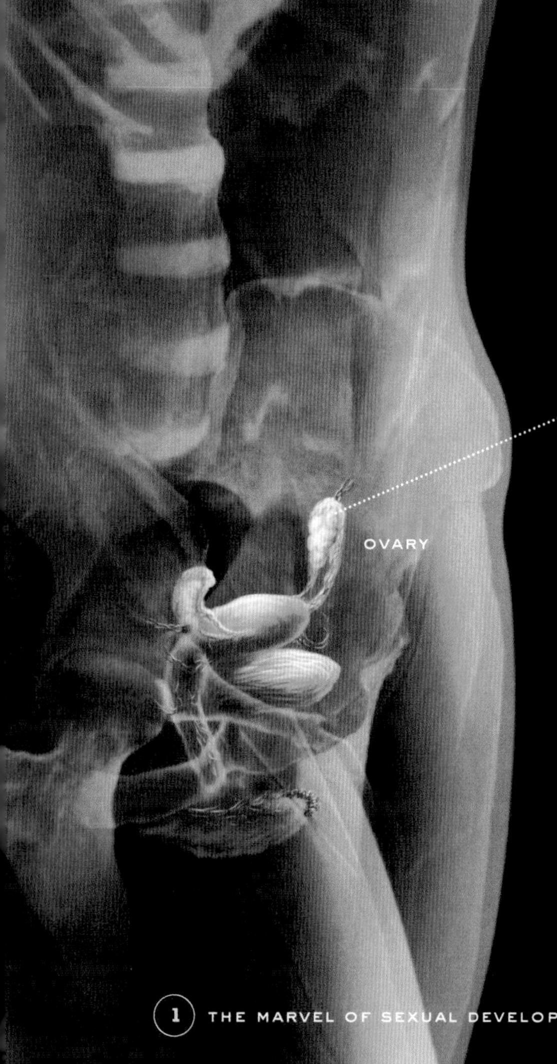

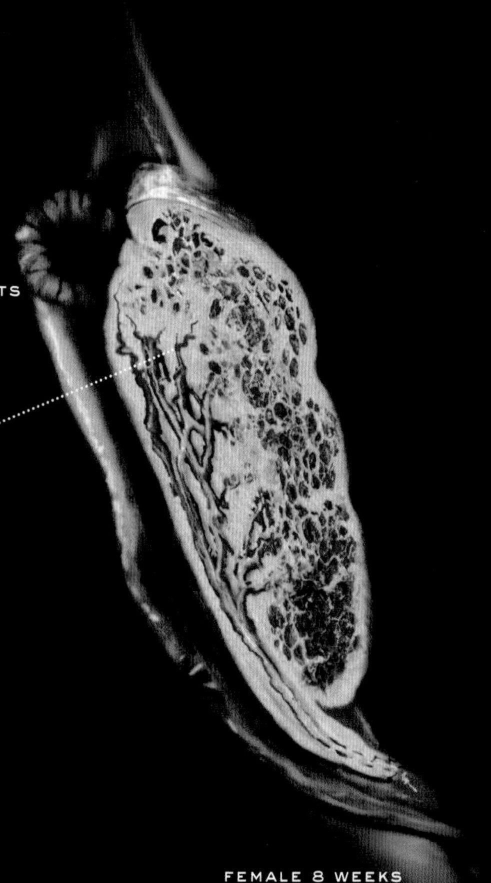

COMPARISON OF THE MALE AND FEMALE UROGENITAL ORGANS DURING THE 8TH WEEK. THE DEVELOPING OVARIAN DUCTS CAN BE CLEARLY SEEN IN THE FEMALE. AND THE DEVELOPING TESTICLES AND SEMINAL VESICLES IN THE MALE.

to grow, XY fetuses would end up both male *and* female on the inside. The Wolffian ducts, on the other hand, will degenerate unless they're stimulated by testosterone to grow into the epididymis, vas deferens, and seminal vesicles.

The testicles start producing AMH and testosterone after they differentiate in week six. If, in a male fetus, the testicles don't form correctly, they may not produce AMH or testosterone. Either mistake can have serious effects. For example, a fetus that produces and responds to AMH, but not testosterone, may have no obvious internal genital development. A fetus that produces or is exposed to testosterone, but doesn't make or respond to AMH, may actually have characteristics of both male and female genitalia. Fortunately, these abnormalities are rare.

Finally, by the 28th week, the testicles have descended to their final location in the scrotum. The inguinal canal, which remains after birth in both men and women, will have been used almost like an elevator—the gubernacula, a cord-like ligament inside it, pulling the testicles down. The ovaries, in contrast, stay where they are in the abdomen—tethered by the body in their well-protected location.

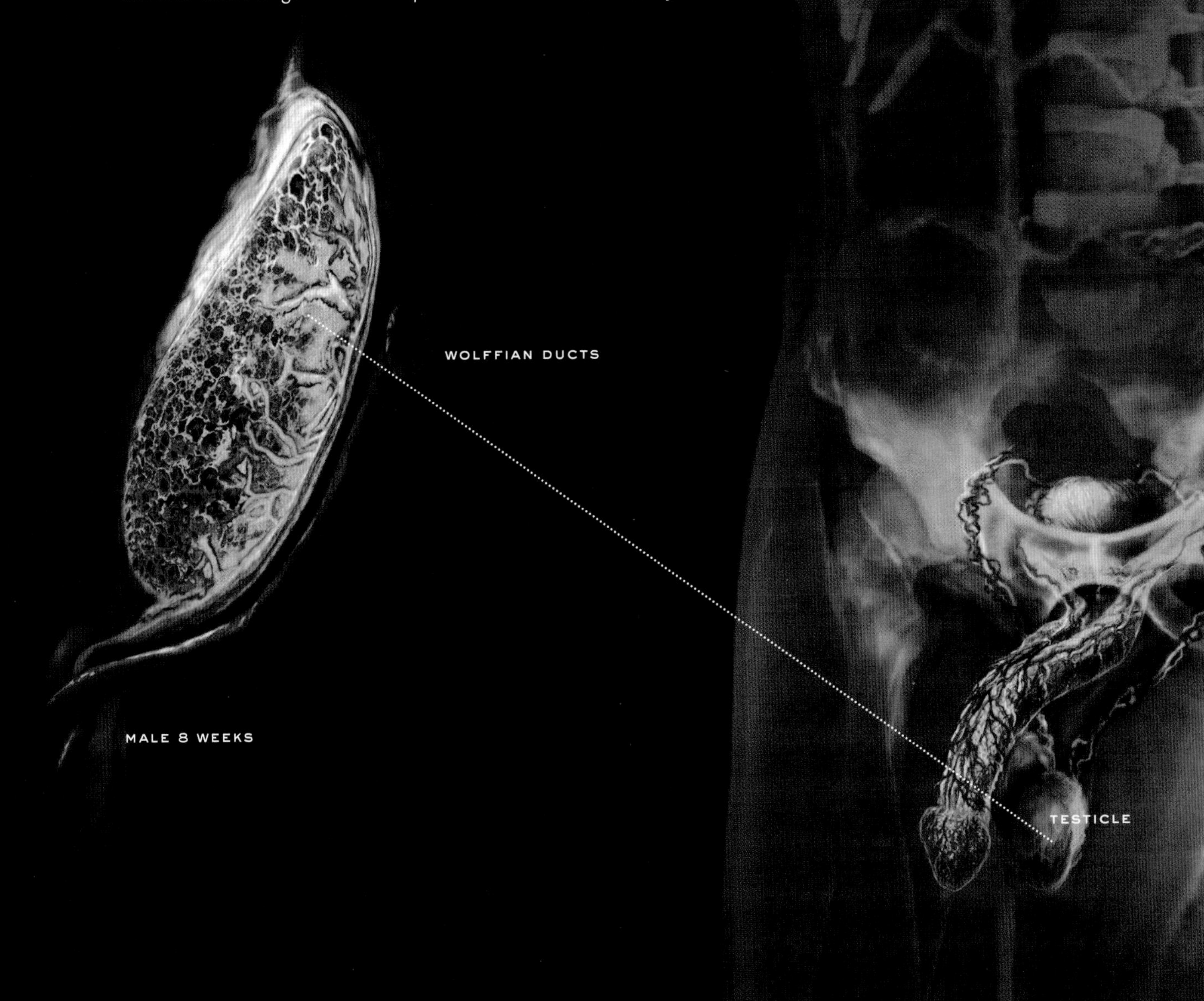

When we look at a child, it's not the differences on the inside that we see, but those on the outside. The irony is that although the *internal* genitalia develop from two separate duct systems, the *external* genitalia of men and women are actually derived from the same tissues! The genital tubercle (which will become the penis or the clitoris), urogenital groove and sinus, and labioscrotal folds appear in all fetuses by the end of the 7th week of development. What they develop into depends on whether or not they're stimulated by the fetal androgens (testosterone and dihydrotestosterone, or DHT). If they're not stimulated by these androgens, they develop into the female sex organs—the clitoris, urethra, vagina, and labia. If they are stimulated by androgens, they merge into the two male sex structures—the penis and scrotum. The external genitalia are, therefore, thought of as *homologous* structures . . . which is just a way of saying that they're really very similar. In a male, what would be the clitoris in a female lengthens into the penis, the urethra becomes enclosed in its shaft, and the labia fold to enclose the testicles as the scrotum.

LOOK! IT'S A. . .

9-WEEK FEMALE

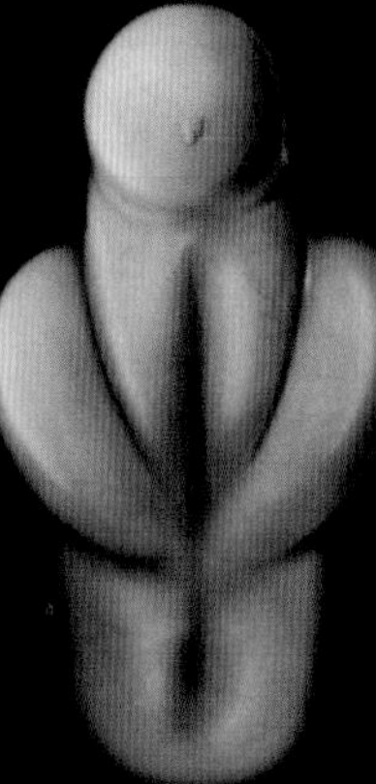

11-WEEK FEMALE

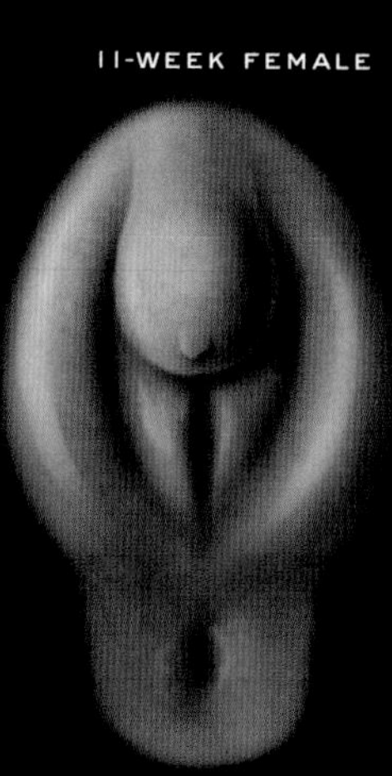

UNDIFFERENTIATED PENIS AT 7 WEEKS

9-WEEK MALE

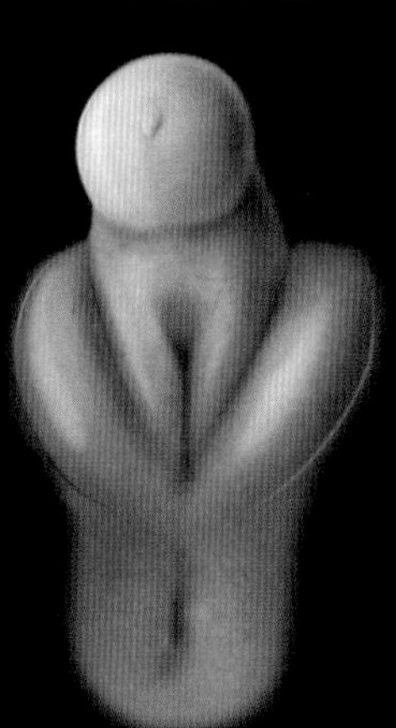

11-WEEK MALE

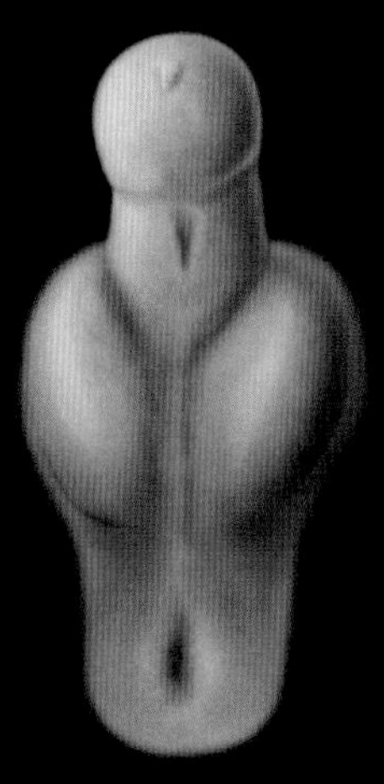

Male fetuses become visibly different from female fetuses in the second and third month of pregnancy. It's not the commonly known hormone testosterone, but instead its highly active derivative DHT, that is primarily responsible for this differentiation. Male fetuses that lack 5-alpha reductase, the enzyme that transforms testosterone into DHT, may be incompletely masculinized and have indeterminate external genitalia that are neither completely masculine nor completely feminine. Similar indeterminate genitalia can be seen in female fetuses that are exposed to androgens while in the womb. Such androgen exposure can be caused by environmental exposures, certain drugs, and maternal or fetal health problems. At puberty, the external genitalia develop further, and in individuals with an abnormal hormone mix the appearance of the genitals may change completely. But in general, adult sexual anatomy is established by the end of the third trimester of pregnancy.

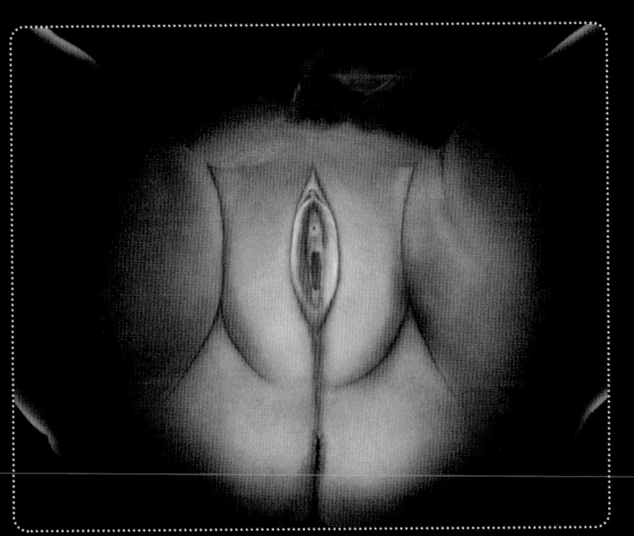

THE CLITORIS, LABIA, AND VAGINA ARE ALL VISIBLE STRUCTURES OF THE EXTERNAL FEMALE GENITALIA AT BIRTH.

NEWBORN

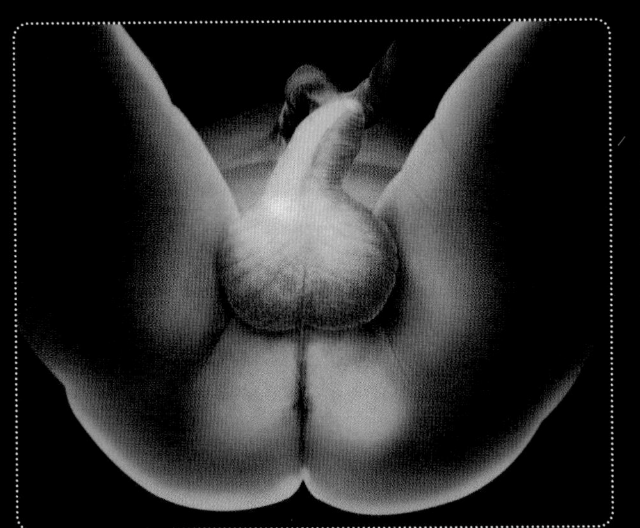

AT THE TIME OF BIRTH, THE MALE GENITALS APPROXIMATE THEIR ADULT APPEARANCE—THE PENIS IS FORMED AND THE TESTICLES HAVE DESCENDED INTO THE SCROTUM.

5-alpha reductase deficiency results from the inability to convert testosterone to dihydrotestosterone. Individuals with this condition are born with ambiguous genitalia and generally presumed to be female at birth. A rather startling transformation occurs around age 12, when the affected individual suddenly finds herself with testicles descending into her scrotum, a clitoris that enlarges into a penis, and a host of secondary sexual characteristics. The condition is particularly frequent in a small village in the Dominican Republic, where the affected individuals are referred to as "guevedoches," which roughly translates to "balls at 12." [1]

WHERE DID THESE COME FROM?

Shopping for the first bra is a pivotal moment at the start of womanhood, but breasts aren't the only things that grow during puberty. In girls, the growth spurt begins before puberty, and actually tapers out at around the time of the first period, which is known as *menarche*, from the Greek words for "moon" (or monthly) and "beginning."

At around the time that breast growth begins, the brain increases its production of the hormones that will regulate a woman's reproductive life. The increase in follicle stimulating hormone (FSH), luteinizing hormone (LH), and gonadotropin-releasing hormone (GnRH) stimulates the development of the ovaries. Estrogen produced by the ovaries causes most of the changes in the external genitalia and also plays an important role in bone growth. Estrogen is also created in fatty tissue, however, and extremely slender young women may have delayed puberty. This is most commonly seen in athletes like dancers and gymnasts, who sometimes maintain a childlike appearance well into their late teens or 20s. A minimum amount of body fat is desirable and even necessary, because too little body fat or too-severe dieting may delay and even prevent menarche. For most young women, it takes almost two years of change for her body to prepare for her first period.

Menarche is not the only "first" for women that involves blood. Blood is also associated with another important moment in a young woman's life: the time of her first sexual intercourse. In fact, not all women bleed at this time, and those who do experience varying amounts of discomfort. In some cultures, the "blood on the sheets" is a traditional way to confirm a young woman's virginity the morning after she is wed. Bleeding during first intercourse is the result of the disruption of the hymen—a piece of membranous tissue that blocks the vagina to varying degrees. The hymen almost always has an opening wide enough to allow for unimpeded menstrual flow or for the insertion of a tampon. But the size of the hymen can range from a thin ring that stretches easily and doesn't impede penetration at all, to a thick covering that actually must be opened by a medical professional to allow intercourse to occur. This condition is called having an imperforate hymen, and it occurs in approximately one in every thousand women.

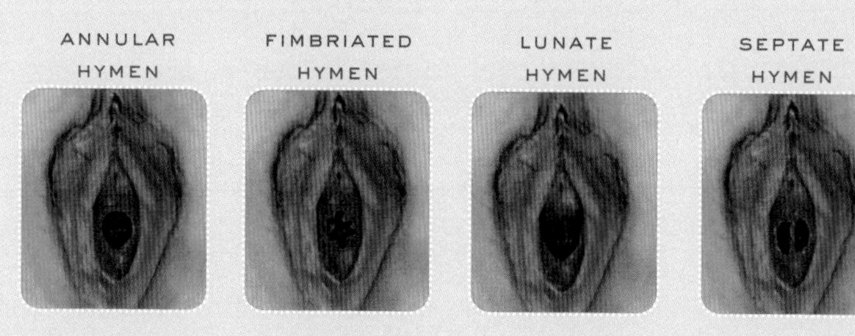

MASTURBATION

Contrary to popular belief, masturbation is not a behavior that begins at puberty. It has been observed in children and infants, and in general it is a healthy behavior that causes no physical or psychological problems. In other words, it's not going to make you go crazy or blind. And, although men are more likely to masturbate than women and tend to masturbate more frequently, it also isn't going to make hair grow on your palms . . . or anywhere else. Since masturbation carries no risk of pregnancy or sexually transmitted disease, some people, including former Surgeon General Joycelyn Elders, consider it to be the safest form of sexual expression. It may also have health benefits: masturbation is known to reduce the severity of menstrual cramps, and one study has suggested that frequent ejaculation during a man's 20s and 30s may actually decrease his risk of developing prostate cancer.

DURING PUBERTY, THE BREASTS BECOME PADDED WITH FAT AND THE MILK DUCTS DEVELOP INTO A FUNCTIONAL FORM.

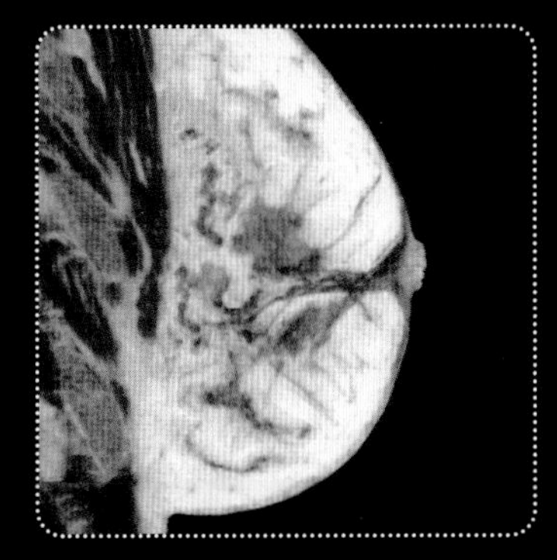

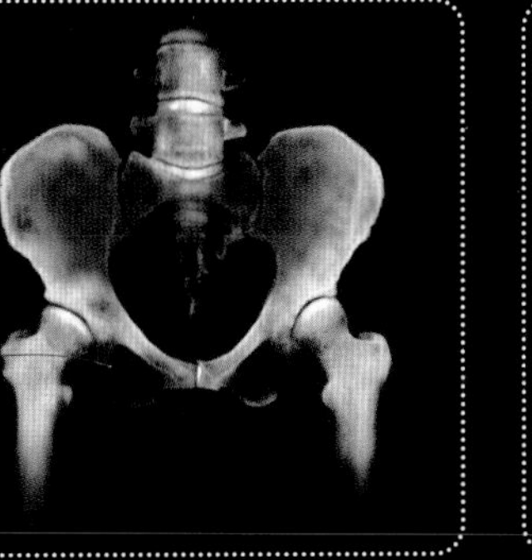

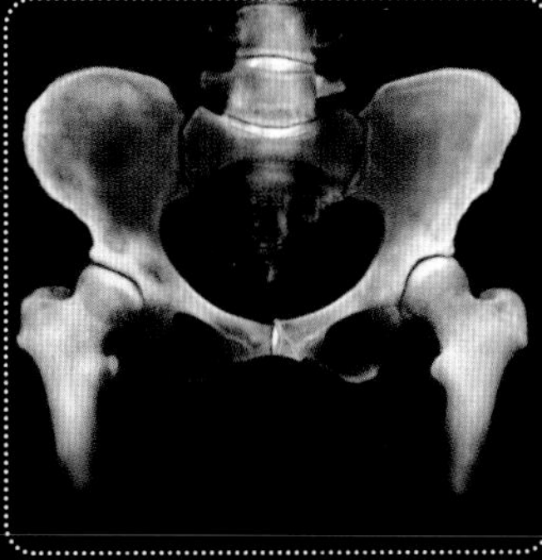

THE BONES OF THE HIPS SPREAD TO

The curse. The gift. The visitor. These are only a few of the words and phrases that people use to describe the shedding of the uterine lining that signals a young woman's body is capable of bearing children. In other words, they're euphemisms for menstruation—and they reflect the wide range of attitudes toward this pivotal monthly event. A woman's first menstrual period may be exciting or exhausting, painful or profound, depending on how and where she was raised. In some cultures it's a source of pride and celebration, while in others it's barely discussed. To various degrees, many women are segregated from their society when they're bleeding: they may be prohibited from prayer, required to abstain from sex, or physically separated from their families during the time of menses. One of the most interesting rituals surrounding menstruation is undertaken by Orthodox Jewish women who follow the rules of *Taharat Hamishpachah*, or family purity. These women partake in a ritual bath, the *mikvah*, after the end of each menstrual period before they can resume sexual relations with their husbands. In recent years, many less religious women have also turned to this activity as a way of regularly reaffirming both their sexuality and their faith.

The uterus is a self-cleaning organ: every month it cycles out the residue of reproduction to start fresh again. During the early phase of the menstrual cycle, rising estrogen levels cause the lining of the uterus to thicken. This rise in estrogen is stimulated by the FSH, which is made in the anterior pituitary gland. The follicle is the receptacle in the ovary in which the egg develops, and the FSH stimulates the egg's growth while it is in the follicle. At the middle of the cycle, there is a surge of another pituitary hormone, the LH. Now ovulation, release of the egg, occurs. At this point, the cells that had been nurturing the egg start producing the sex hormone progesterone, which causes the uterine lining to further thicken and prepare for pregnancy. If pregnancy does not occur, estrogen and progesterone levels fall and the uterus sheds its thick-

GO WITH THE FLOW

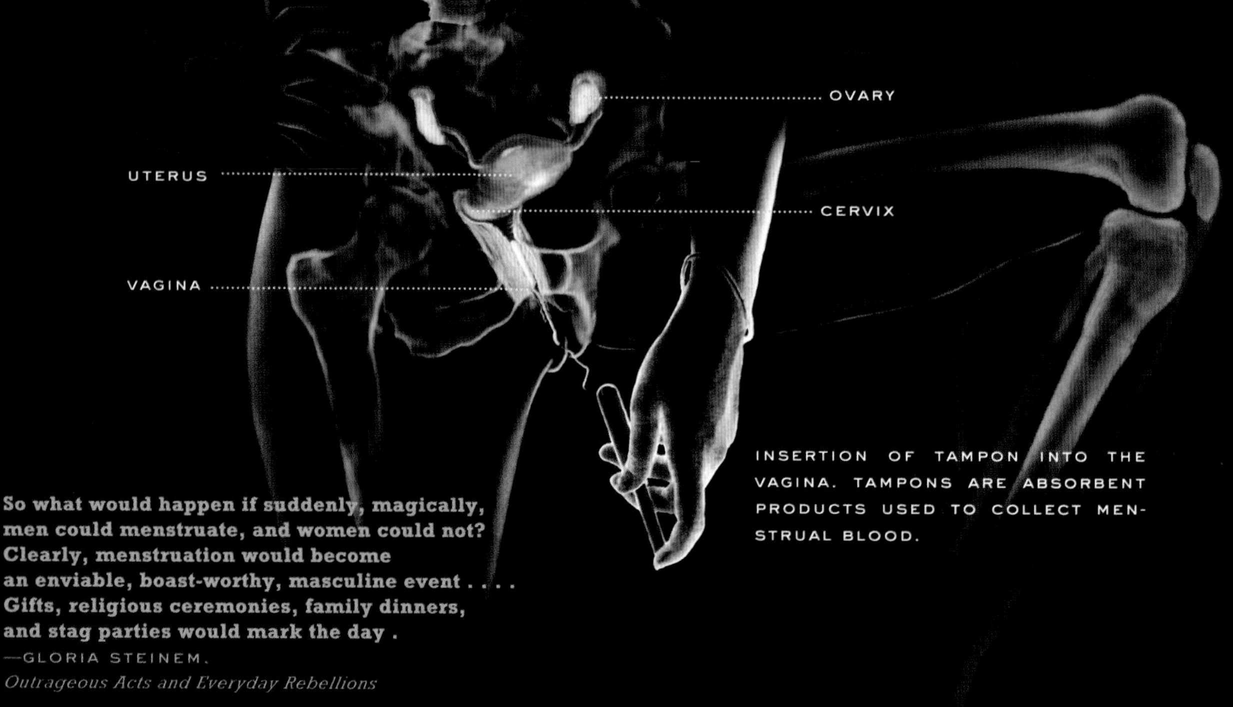

INSERTION OF TAMPON INTO THE VAGINA. TAMPONS ARE ABSORBENT PRODUCTS USED TO COLLECT MENSTRUAL BLOOD.

So what would happen if suddenly, magically, men could menstruate, and women could not? Clearly, menstruation would become an enviable, boast-worthy, masculine event . . . . Gifts, religious ceremonies, family dinners, and stag parties would mark the day .
—GLORIA STEINEM,
Outrageous Acts and Everyday Rebellions

ened lining to prepare for another reproductive cycle. This shedding, menstruation, usually lasts three to seven days and removes both blood and tissue from the uterus.

Sometimes menstruation can be a very unpleasant process. The hormones involved, and the process of expelling blood and tissue from the uterus, can cause emotional distress, cramping, stomach pains, diarrhea, breast tenderness, and numerous other symptoms. Around 15 percent of women have pain so severe during their periods that it's actually disabling and affects their ability to go to school or work. A smaller percentage have mood swings so serious that they consider suicide. Fortunately, most women's periods cause only a moderate amount of discomfort.

Just like the uterus, the vagina is also self-cleaning. Vaginal secretions clean out bacteria and debris, and the vaginal interior does not need to be washed. Douching is, in fact, a major source of vaginal and reproductive health problems, not a cure for them. Instead of cleaning the vagina, douching is far more likely to wash bacteria and other organisms up into the uterus where they can change from being a minor irritation to a source of serious complication.

Menstruation is not only an important part of a woman's life, it's also the basis for a major industry. Most woman menstruate every month that they aren't pregnant, and they do so for more than 20 years. Over this time women use a huge number of menstrual collection devices. Although the most common products work by absorbing blood either externally (menstrual pads) or internally (tampons), a growing number of women are experimenting with different types of menstrual collection devices. Reusable menstrual cups or pads, for example, are appealing to women who dislike the waste associated with disposable products. Menstrual cups are also good for women who enjoy having sex during their periods, because unlike tampons or pads they can be used to block the flow of menstrual blood during penetration.

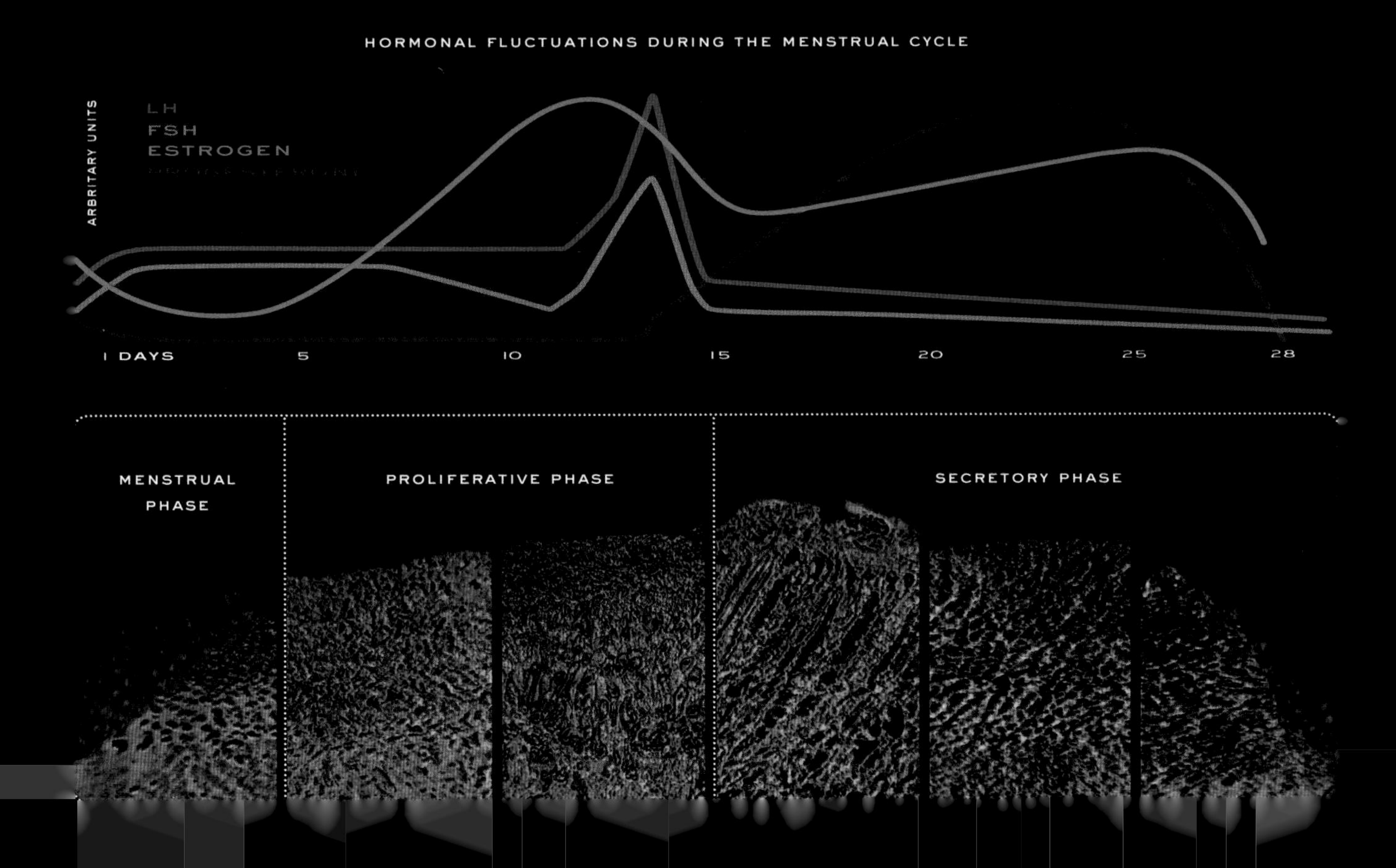

SIZE DOESN'T MATTER

In men, unlike in women, the onset of puberty is not visible to the outside world. Instead, the first sign that adulthood is imminent is the growth of the testicles, which increase in size as their hormone production begins to increase. These hormones, in turn, cause the growth of pubic hair, the thinning of the scrotum, and the enlargement of the penis. The locker room question, though, of how large a boy's penis is will rarely be answered before he turns 16, and some boys maintain penis growth for several more years. As puberty continues, boys' muscle structure increases, their voice "breaks," and they finally start to grow—after several uncomfortable years of being shorter than most of the girls in their grade. Although most men are taller than most women, it's estrogen that is in charge of bone growth. Boys don't have enough estrogen in their system to start their growth spurt until girls are already reaching their final height.

Puberty in males can be an audible experience. Although vocal changes occur in girls as well, these changes are much more noticeable in boys. In boys, when the larynx increases in size, it also experiences a growth of cartilage into what is commonly called the Adam's apple. As the voice deepens and the vocal cords lengthen, these changes can cause some loss of vocal agility. This "cracking" has been the source of humor in many teen movies and television shows. Fortunately, it resolves over time.

Although it has profound effects on fetal sexual development, during puberty the effects of DHT are only slightly stronger than those of testosterone. The large volumes of testosterone that are suddenly present more than make up for the strength of DHT, which is why some genital abnormalities may actually resolve at this time. Genetically male children lacking in DHT may appear to be female until puberty, at which point the increase in testosterone causes the testicles to descend and the apparent clitoris to enlarge into a penis.

It's not just a boy's body that changes during puberty; so does his brain. Although men have the same regulatory hormones as women (GnRH, LH, and FSH), their brains express them in a completely different way. Instead of varying monthly to create a menstrual cycle, the hormones in men's brains are constantly active. Maybe men and women really do have different ways of thinking about sex. . . .

MASTURBATION

Hey, don't knock masturbation.
It's sex with someone I love.
—WOODY ALLEN *as Alvy Singer in Annie Hall (1977)*

Just as a girl's first period is a defining moment of womanhood, a boy's first nocturnal emission, or wet dream, is a defining moment of manhood. It happens when hormones have reached new, higher levels in a young man's system. The increase in sperm production and sexual tension can cause frequent involuntary erections, both at night and during the day. The three-ring binder might have been invented as a discreet way for young men to hide an unfortunate erection occurring during the school day! Masturbation is a natural response to this increased level of sexual arousal. For teenage boys, it is an instinctive and effective way to reduce the number of spontaneous erections and the embarrassment they sometimes cause.

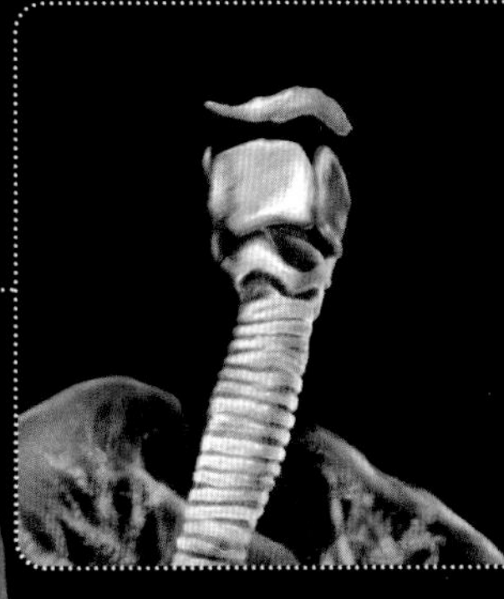
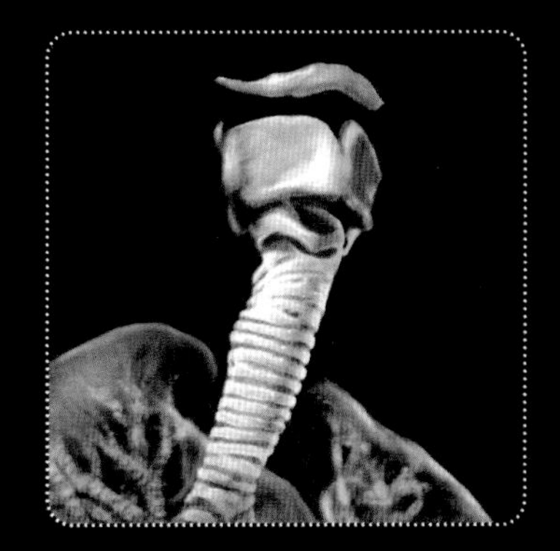

IN MALES DURING PUBERTY, TESTOSTERONE CAUSES THE LARYNX, OR VOICE BOX, TO INCREASE IN SIZE.

The voice deepens during puberty because testosterone causes the larynx, or voice box, to increase in size. The vocal cords also lengthen at this time. Some boys may experience their voice "cracking" in this period, jumping back and forth between different types of sounds, as their body tries to adjust to the changing shape of their vocal organs. In the not-so-distant past, boy sopranos with beautiful voices were castrated to keep their voices intact.

Shaving for the first time can be a very exciting experience for boys. During puberty, facial hair begins to grow on the face. It also begins to grow on the underarms and legs, in the pubic area, and on the chest. Not all men develop chest hair, and for some men it begins to grow much later in life. Chest hair, like acne, is highly affected by a man's levels of dihydrotestosterone, which can change over time.

Body shape changes dramatically as boys grow into manhood. Height increases, the shoulders broaden, and muscle growth is vastly increased. At this time, there may be some swelling of a boy's nipples, but this is normal and does not mean that he is growing breasts.

Penis and testicular growth can be a stressful component of puberty. Boys can feel compelled to compare their penis size to that of their friends, but it is important to remember that size has nothing to do with masculinity. And, although in most cases the penis has reached its final size by age 16, in some men it can continue to grow for several more years.

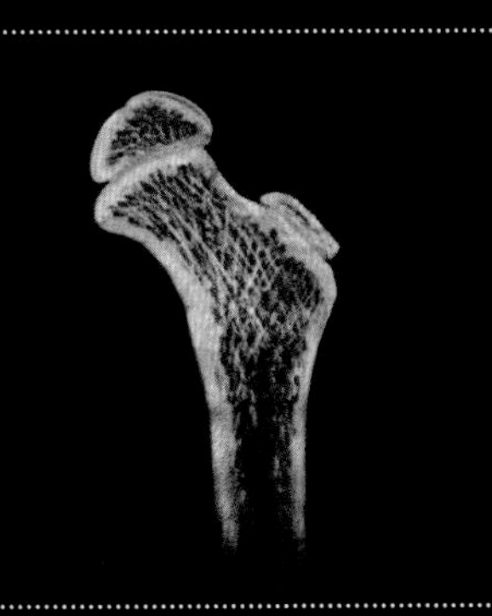
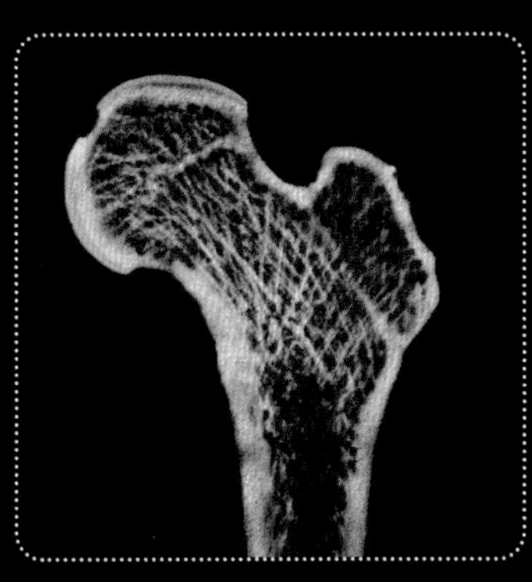

AS PUBERTY PROGRESSES, THE LONG BONES OF THE BODY GROW AND MATURE, AND THE ADOLESCENT INCREASES IN HEIGHT.

CROSS SECTION OF A SEMI-NIFEROUS TUBULE, THE SITE OF SPERM DEVELOPMENT AND TESTOSTERONE PRODUCTION

CROSS SECTION OF A TESTICLE

TESTOSTERONE MOLECULE

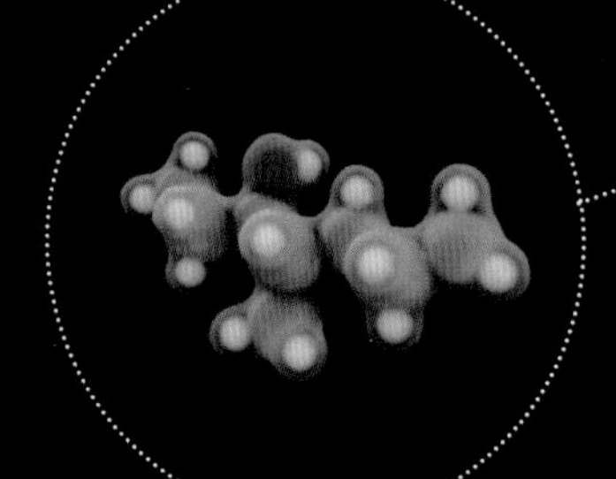

SPERM AND TESTOSTERONE: THE NEXT GENERATION

TESTOSTERONE PRODUCTION Just as the pituitary hormones FSH and LH are important in establishing the menstrual cycle, they also have important effects on male hormone production. LH stimulates the growth of the leydig cells, which are the primary site of testosterone production. FSH is far more involved in the stimulation of sertoli cells, which are involved in the formation of sperm, or spermatogenesis.

Testosterone is mostly produced in the leydig cells, but low levels are also synthesized in the adrenal cortex. Testosterone starts out as cholesterol and then goes through four steps before reaching its final form. Leydig cells are found in the body of the testicles, along with blood vessels and macrophages, not in the seminiferous tubules where sperm are produced. Leydig cells are vital to spermatogenesis since without their testosterone production it would not occur. That the level of sperm production is directly proportional to the number of leydig cells further shows their importance.

SPERMATOGENESIS The production of sperm takes almost three months from the first meiotic cell division until the fully grown sperm are ready for ejaculation. Spermatogenesis is near its peak in the late teens and early twenties. Unlike women, who are born with a fixed number of gametes, or reproductive cells, that slowly are used or die off over time, men are constantly making new sperm. This means that men may remain fertile throughout their lifespan, although as they age the level of sperm production may drop. This decline can be due to decreased testosterone levels or to loss of elasticity, or *sclerosis*, in the seminiferous tubules. Although the prostate makes important contributions to the volume of seminal fluid, it is possible for a man without one to remain fertile.

FIVE STAGES OF PUBERTY The progress of puberty can be assessed using the Tanner Stages of Puberty Scale, developed by the British physician Jim Tanner. During stage one, boys grow slowly, around two inches per year; they have no coarse pubic hair and no penile growth. During stage two, growth remains similar, but pubic hair begins to appear and the penis begins to increase in both length and width. At this time, fat levels also begin to decrease and the scrotum thins and reddens. During stage three, growth increases to three inches per year, pubic hair is found all over the pubis, and the voice may break. It is also during this stage that muscle mass begins to increase and there may be some swelling behind the nipples. During stage four, growth is at its fastest rate of around four inches per year, underarm hair appears, vocal changes continue, and acne begins to be a problem for some young men. Finally, in stage five, growth of both the body and the penis ceases, facial hair appears along with a mature male physique, and pubic hair spreads to the thighs. Although there are no set durations for any of the stages, they can be used to monitor a teenager's progression to adulthood and may help to flag any abnormalities in growth or hormonal development while they are still easy to treat.

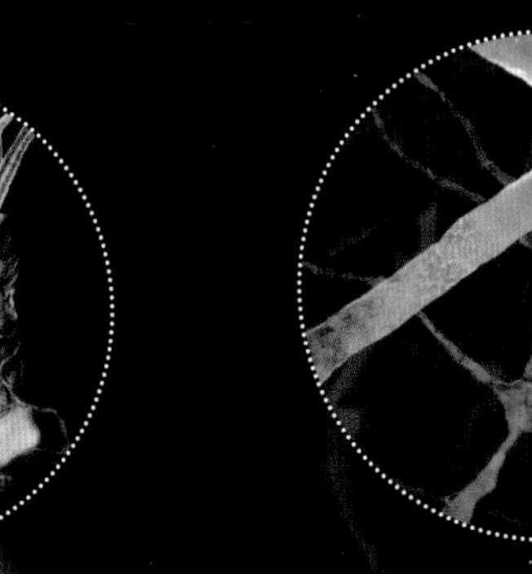

FAT CELLS WITHIN THE PUBIC REGION PRODUCE ESTROGEN.

THE NEURONS THAT CONVERT TESTOSTERONE TO ESTROGENS ARE IN THE BRAIN.

THE HYPOTHALAMUS AND PITUITARY GLAND SIGNAL VARIOUS STRUCTURES TO PRODUCE THREE ESSENTIAL SEX HORMONES: ESTROGEN, TESTOSTERONE, AND DHT. THESE HORMONES HAVE MULTIPLE PHYSIOLOGICAL EFFECTS RANGING FROM INFLUENCING GROWTH, SEXUAL FUNCTION, AND METABOLISM.

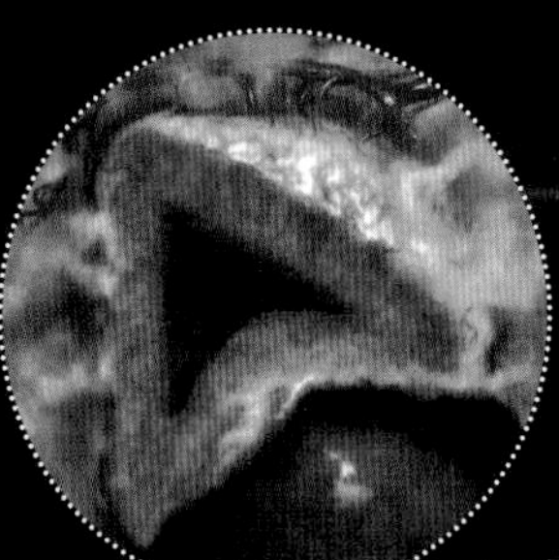

THE ADRENAL GLANDS PRODUCE SMALL AMOUNTS OF TESTOSTERONE AND ESTROGEN.

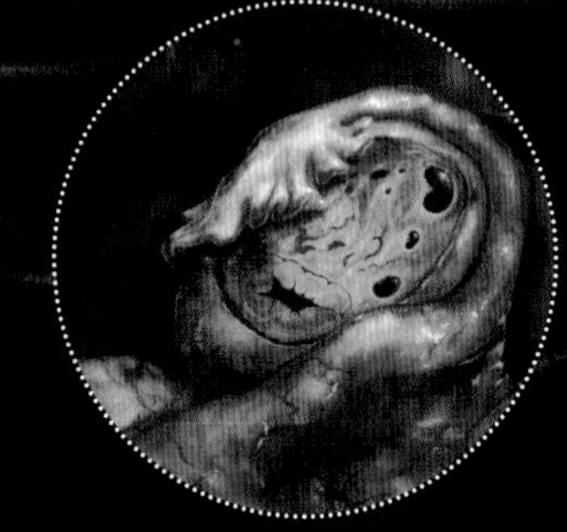

THE OVARIES PRODUCE ESTROGENS AND PROGESTERONE.

HORMONES IN CHARGE

Sexual differentiation isn't a process that only happens in the body—it also happens in the brain. Exposure to hormones, like testosterone and DHT, affects the way the brain develops during pregnancy as well as a person's behavior after birth. Although the genes that control their development are largely the same, male and female brains are functionally quite different. For example, women are better at distinguishing colors, while men are better at tracking motion. There are also many subtle developmental differences that affect both sex and sexuality. A great deal of research suggests that hormone exposure before birth shapes sexual identity throughout life.

The three main sex hormones are considered to be estrogen, testosterone, and DHT. But despite their designation as sex hormones, their roles are not confined to the genital tract. They're not even made just in the gonads, but in the skin, adrenals, and other organs as well. One of the main sites of estrogen production is actually the fatty tissue. This is why weight can have such a profound effect on the timing of puberty. The effects of sex hormones aren't limited to reproduction; they work on many of the body's systems. Estrogen, for instance, affects bone growth, and testosterone and DHT influence the formation of hair and muscle.

One of the most interesting things about sex hormones is that, contrary to popular belief, all three hormones are present in both sexes. Estrogen is not just a female hormone, any more than testosterone just belongs to males. Yes, women have more estrogen than men, but some research suggests that estrogen may be essential for maintenance of the male libido, or sex drive. Men have greater levels of testosterone, but without enough of it sexual desire and function in women is greatly reduced. Some scientists think that it's actually the asymmetrical levels of hormones that create healthy sexual responses in women and men.

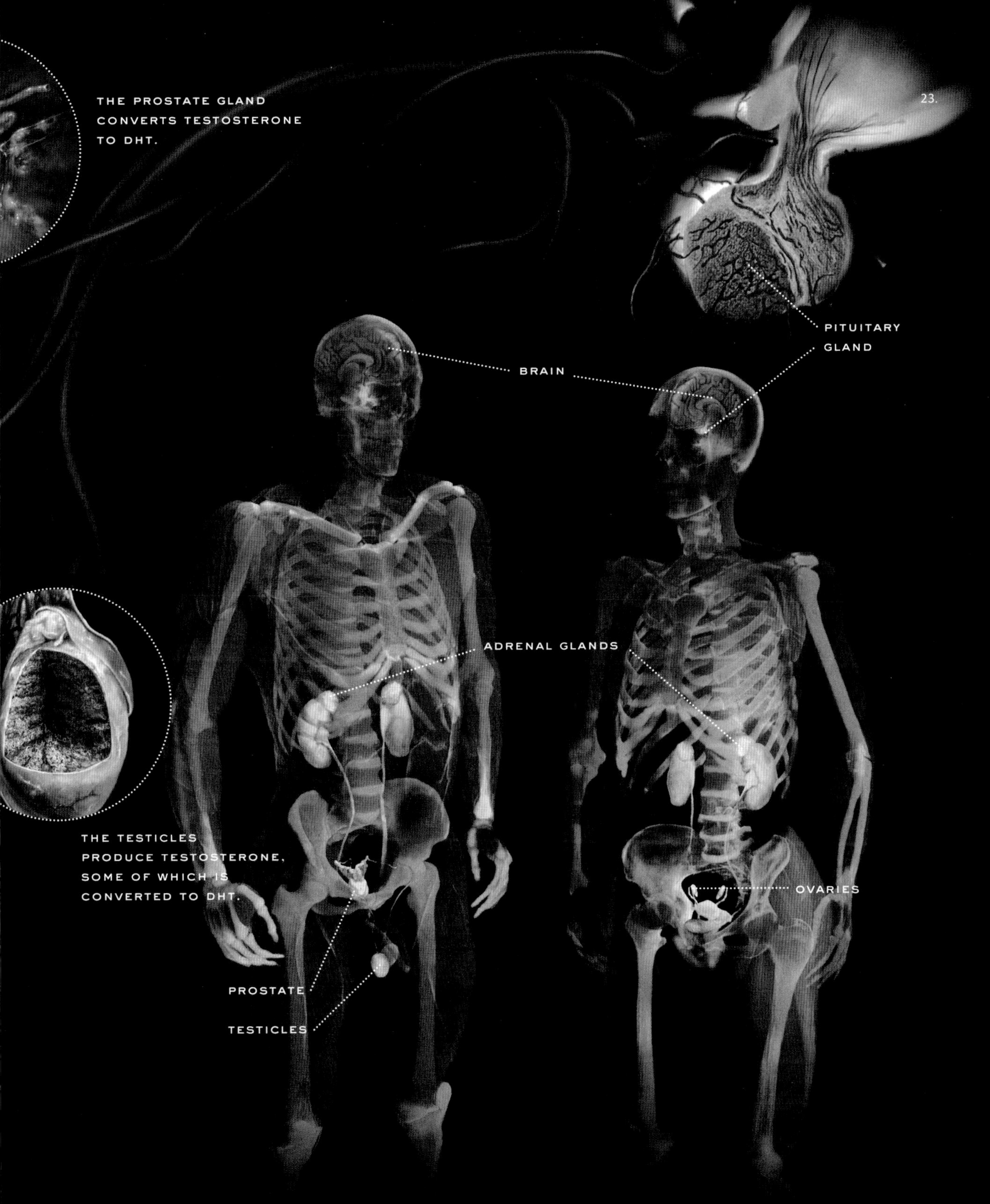

High prenatal exposure to testosterone is associated with having a longer ring finger than index finger. Lower prenatal exposure to testosterone is associated with having ring and index fingers of approximately the same length. This seemingly minor difference can provide cues to a person's health, fertility, sexual orientation, and competitiveness. [2]

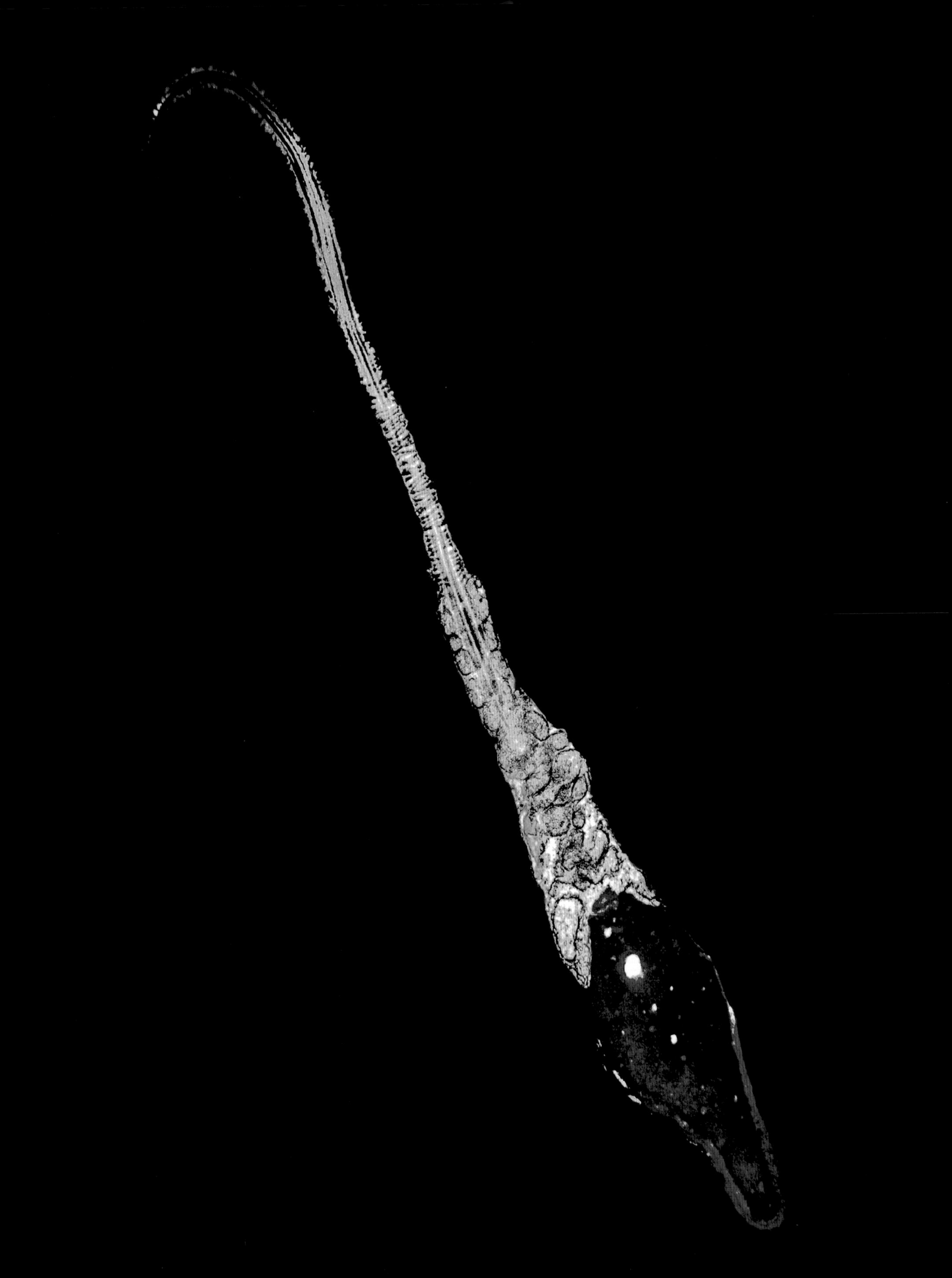

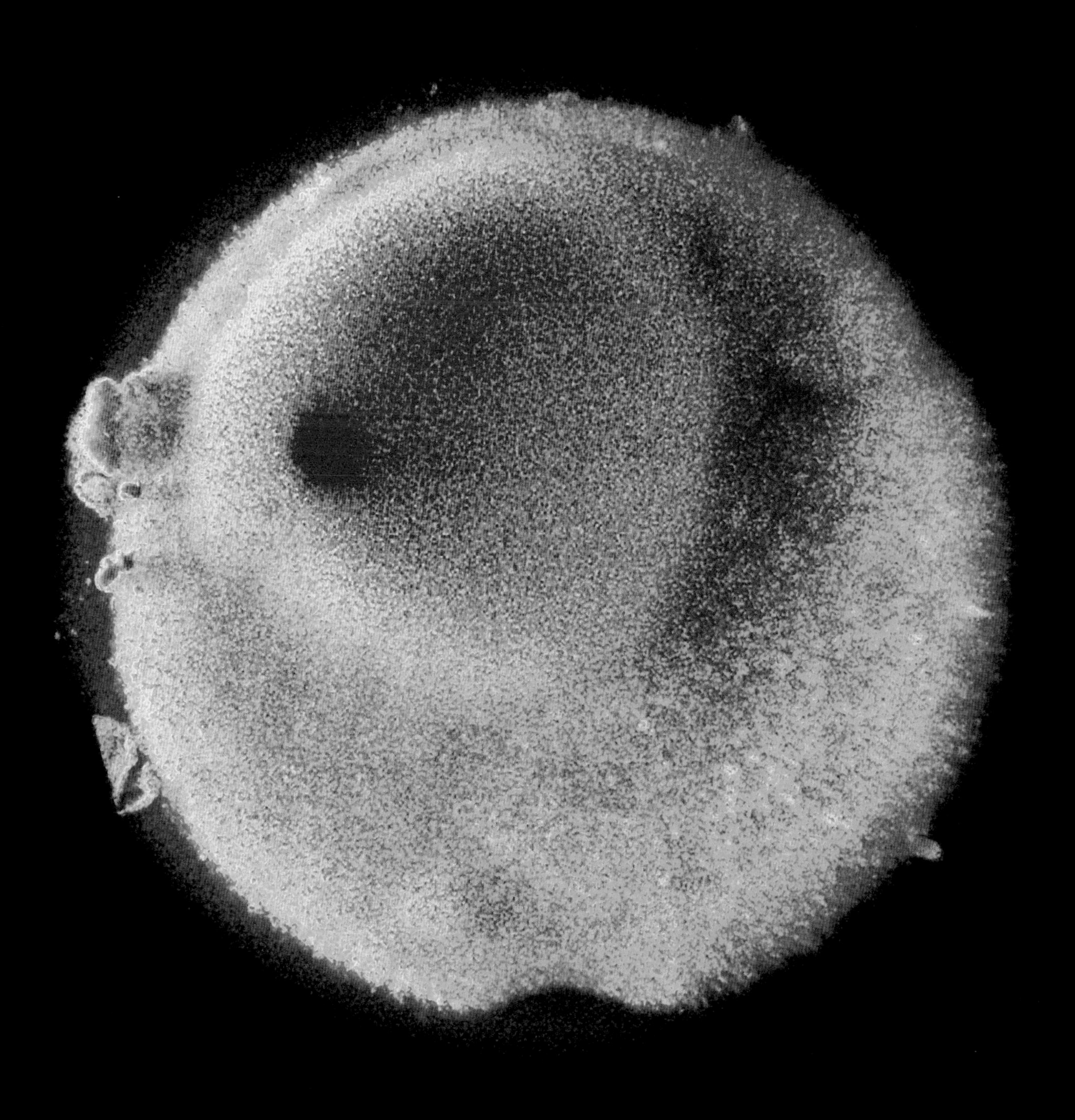

Seeing, hearing, feeling,
are miracles,
and each part and tag of me
is a miracle.

Divine am I inside and out,
and I make holy whatever I touch
or am touch'd from . . .

—WALT WHITMAN, *Song of Myself*

2

SEX: A LOOK BEHIND THE SCENES

The sound of quickened breath, the feel of your heart pounding, and every inch of your skin crying out for a caress . . . now you're thinking about sex. Two people entwining in an act of passionate consummation is more than just exciting; it's also an amazing feat of biological engineering. The changes your body undergoes during a sexual act are astounding, complex, and profound. Are you breathless with . . . anticipation?

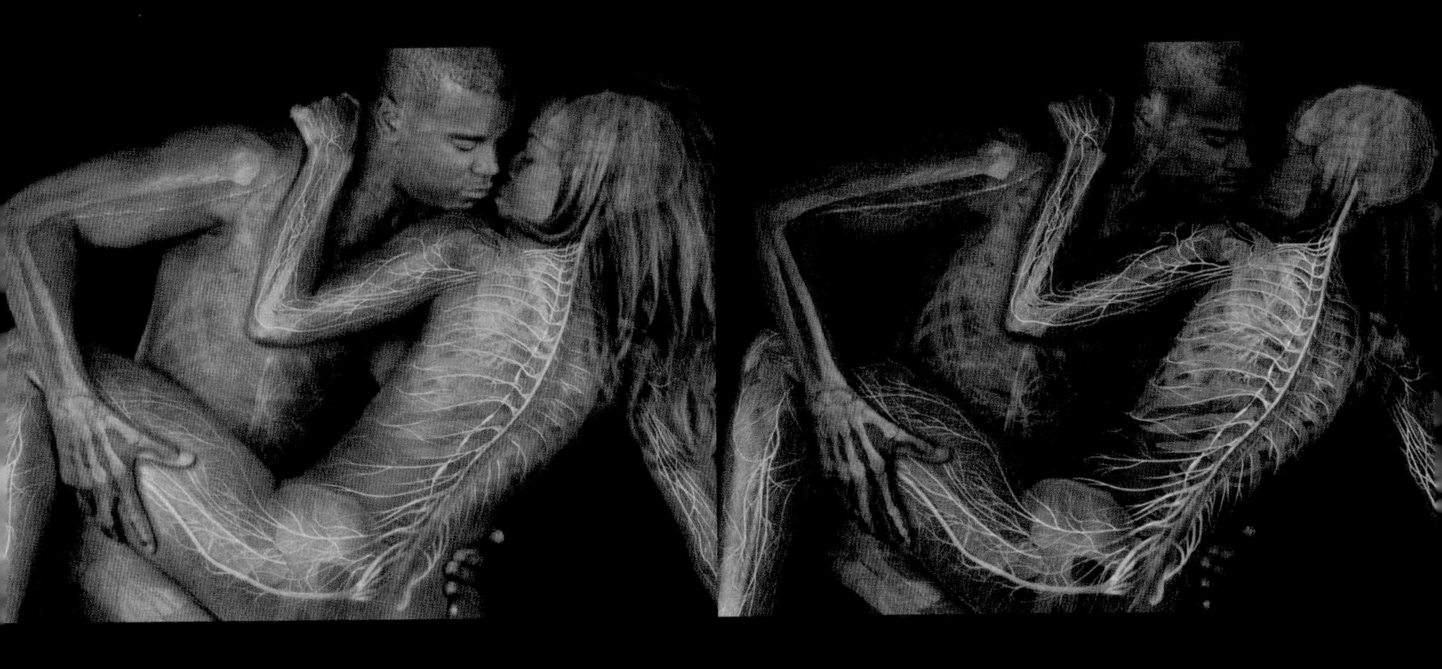

LET'S HAVE A CONVERSATION

ANTICI. . .WAIT FOR IT . . .PATION: Sometimes, there is nothing so enticing as the moment before the first touch. Even then, before anything begins, your body is readying itself for pleasure.

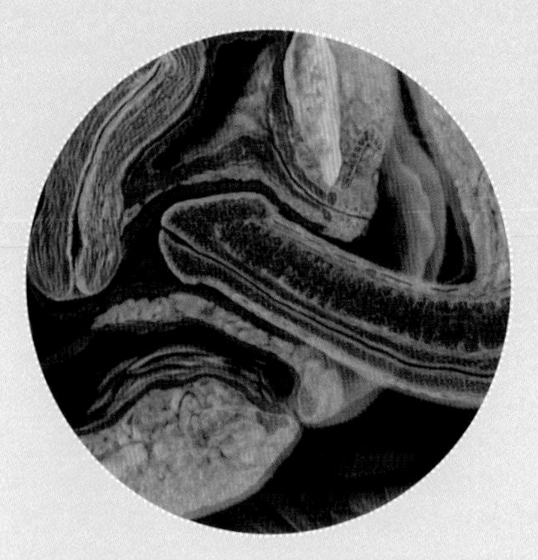

ENJOY THE RIDE: Getting there is more than half the fun. The changes that occur in your body during sexual arousal make every sensation more exciting, even as they urge you towards your passion's goal.

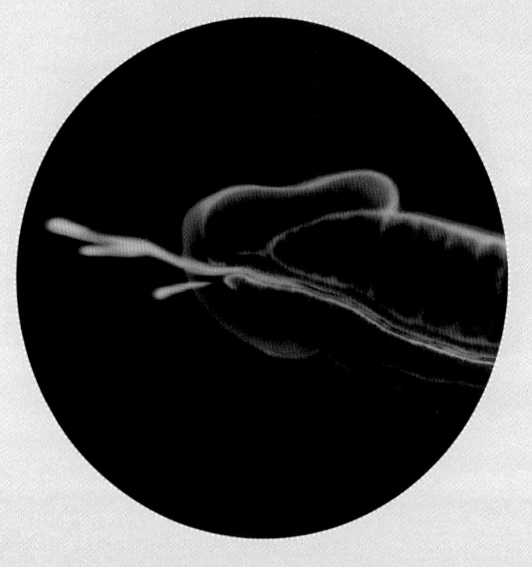

SAVOR THE MOMENT: Orgasm isn't only a pinnacle of sensation; it's also a profound biological release.

THE DESSERT SPOON: In the minutes after orgasm, lying sated in your lover's arms, your body is returning slowly to a tranquil state. A few touches, however, or erotic words, and you may be ready once more for something sweet.

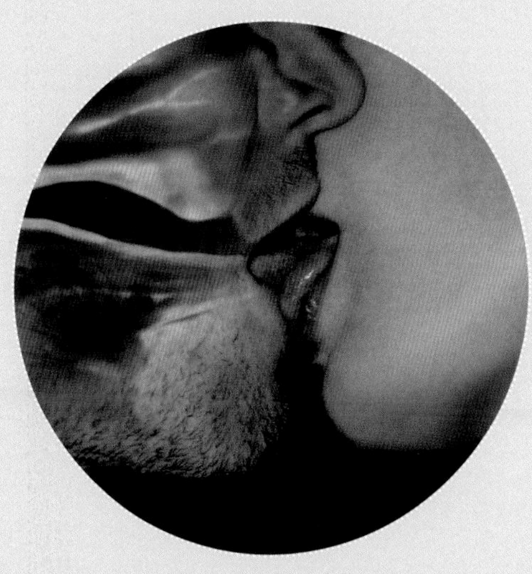

TO KISS OR NOT TO KISS: A kiss is one of the most intimate expressions of sexual desire and affection—or is it? Some cultures believe the mouth should just be used for food.

THE MORNING AFTER: People do the dumbest things in the pursuit of sex. They make decisions they regret and engage in actions even they themselves can't understand. Why is it, for so many people, that when the groin turns on, the brain turns off?

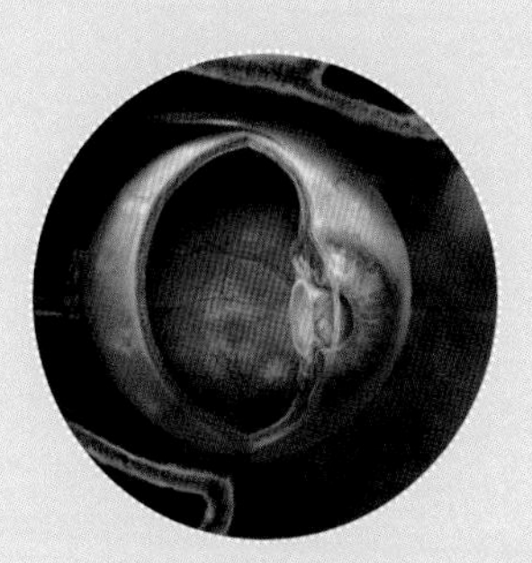

THE EYE OF THE BEHOLDER: Beauty can be arousing in all its forms. A lock of hair, a lushly contoured hip or thigh . . . every person needs a different sight to set their heart ablaze. And the words that incite passion are equally infinite in their variety.

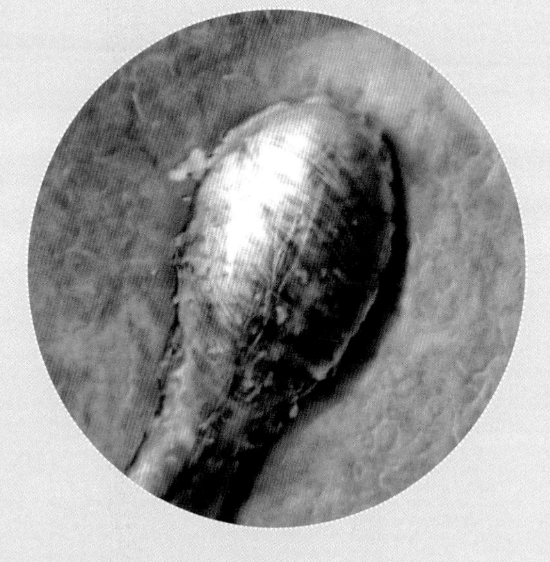

A WORLDWIDE SENSATION: The back of the neck, the sole of the foot, even the dimple of the navel are all magnets for sensual touch. Erogenous zones are known for their concentration of sensitive nerves, but depending on where you live you'll look for them in very different locations. As the translator in the John Patrick play *The Teahouse of the August Moon* says, "Pornography is a matter of geography". . . and so are most of our other ideas about sex.

THE QUANTIFICATION OF DESIRE

The human body is designed to enjoy sex. Whether or not humans evolved this way to increase pair bonding, using sex as a lure to keep men engaged in childrearing, or to improve the quality of the gene pool may never be known. What is clear, though, is that our bodies have developed an enjoyment of an exquisite range of sensations that encourage us to explore the infinite varieties of the sexual act.

No matter in what ways we choose to enjoy sex, our bodies are well equipped for a wide range of sexual experience. The male penis, which is actually longer than most people think since it extends well into the body, is an impressive structure with an extremely sensitive head and a highly developed network of erectile tissue. It is connected to the testicles inside the scrotum, which provide both sexual motivation and material by the production of testosterone and sperm. The testicles hang outside the body because sperm can be adversely affected by body temperature. Located not too far away is the prostate gland, which provides some of the fluid to nourish the sperm and lubricate the mechanisms of intercourse. The prostate gland is not only a functional workhorse, but it is also filled with nerves that, when stimulated through the anus, can be a source of intense sexual pleasure.

A woman's sexual organs are equally complex, and normally hidden inside the folds of the labia majora and minora, the outer and inner lips. As she becomes sexually excited, however, the lips swell and separate to reveal the erect clitoris and the entrance to the vagina. The clitoris is very similar in structure to a small penis, with erectile tissue and a dense concentration of nerve fibers at the head. The vagina is the entrance to the uterus, where during intercourse sperm may be deposited. Even though women have no prostate gland (the Skene's glands are the homologous female structure), they have sensitive nerves in the anus, which can lead to pleasurable feelings during stimulation.

Sex has not always been a subject for science. For most of western history, this act was thought to be too private to be studied and examined objectively. Although there were many scientists who researched sex in a qualitative manner, asking questions about people's experiences and their observations, two scientists named William Masters and Virginia Johnson were some of the first to quantitatively examine sexual physiology. Following the lead of Alfred Kinsey's behavioral studies, they developed experiments and instruments that could actually measure the biological changes during sex, and in doing so were the first to assess the nuts and bolts of physical pleasure. It is for this reason that their four-stage model of sexual desire is so widely accepted: they measured and quantified the sexual response, and used scientific reasoning to break it down into the components of *excitement*, *plateau*, *orgasm*, and *resolution*.

DURING AROUSAL, THE PENIS BECOMES ERECT AND THE TESTICLES ARE PULLED CLOSER TO THE BODY IN PREPARATION FOR EJACULATION.

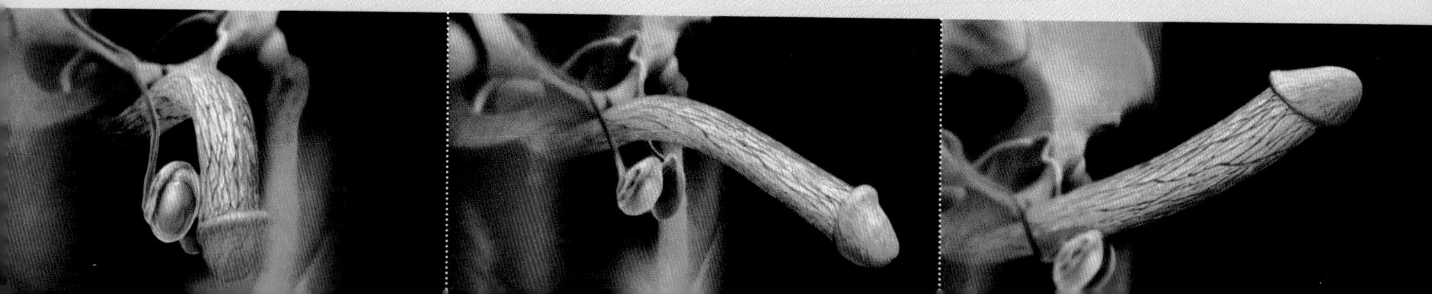

Excitement. It's the first phase of sexual response, where interest turns into arousal and the body begins to respond to stimulus and attraction. As it begins, everything inside you changes. Blood begins to flow to sensitive areas, muscles tighten in anticipation, and your heart begins to race. Although it results in many different sensations, excitement is basically the sum of only two different biological processes. The first is vasocongestion, the dilation of vessels so that blood can flow into the areas where it's most needed. Vasocongestion in the corpora cavernosa in the male and female erectile tissue, and the corpus spongiosum in the male leads to engorgement of the external genitalia. Vaginal lubrication is also the result of vasocongestion. As blood enters the vaginal walls, an increase in fluids seeping through makes a woman wet.

Muscle contraction is the second process that contributes heavily to the biology of sexual excitement. In both men and women, the nipples become erect and more sensitive. In women, the inner two thirds of the vagina expand away from each other like an inflated balloon, providing more room for penetration. Some studies have suggested that the cervix and uterus pull up to allow for easier passage of sperm. In men, the scrotal sac tenses and is pulled closer to the body, while the testicles are tugged in by the shortening of the spermatic cords. Excitement is the body's way of drawing in to prepare for release.

The heady euphoria associated with initial infatuation is associated with increased levels of phenylethylamine (PEA), a biochemical substance similar to amphetamines. [3]

DURING AROUSAL, THE CLITORIS BECOMES ERECT AND THE LIPS OF THE VAGINA SWELL AND SEPARATE.

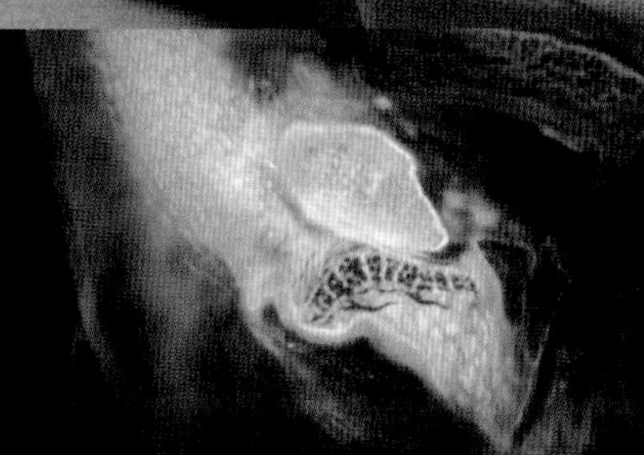

NIGHT PLEASURES

Plateau is the slow uphill climb of the sexual cycle, when the initial sensations of excitement expand luxuriously in preparation for the explosion of orgasm. It is an extended time of building pressure, increasing tension, and heavy breathing. During the plateau phase vasocongestion is at its peak. The penis becomes fully erect, the glans swells, and pre-ejaculatory fluid, secreted by the Cowper's glands, appears at the tip. In women, a structure known as the orgasmic platform begins to grow. The outer third of the vagina tightens, increasing the sensation of penetration, and the clitoris actually draws back into the body as tension continues to increase in the journey toward orgasm.

Plateau is where we spend most of our time during the sexual journey. Every sensation is enhanced in a body that is prepared to make the most out of each moment of stimulation. Traveling toward release can be the most pleasurable part of the sexual experience, and it can be drawn out into hours of enjoyment.

A great deal of the sexual experience actually takes place within the brain. Functional brain imaging has shown that widespread areas are active during sexual response, not just the small regions that were previously known to be involved. These areas are normally activated by genital stimulation, but in some individuals they can be stimulated by thought alone. Some women can experience arousal to orgasm without a single touch. While the genital sensory regions of the brain in the thalamus and cortex are not activated, the orgasm-related brain regions are. This suggests that human cognitive processes can bypass sensory genital pathways and still generate orgasm just by thinking or even during sleep. This may seem surprising, but remember, men have been talking about their nocturnal emissions for hundreds of years; only recently have scientists been able to scan women's brains for evidence of similar activity.

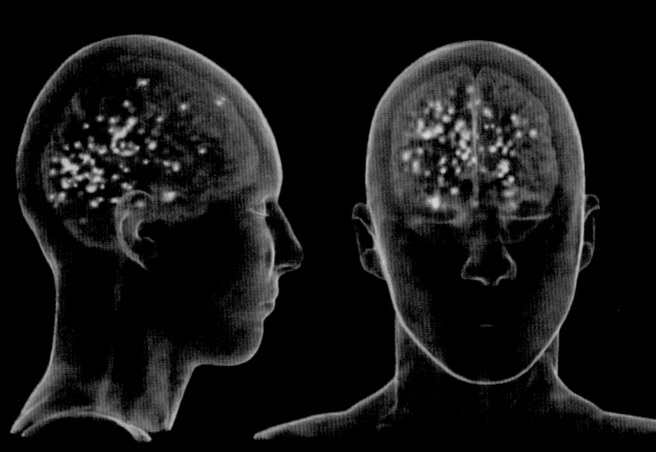

FUNCTIONAL MAGNETIC RESONANCE IMAGING (FMRI) SCAN SHOWS THAT ACTIVITY IS PRESENT THROUGHOUT THIS WOMAN'S BRAIN AS SHE BRINGS HERSELF TO ORGASM THROUGH THOUGHT AND FANTASY ALONE.

Orgasm is, for many people, the goal of the sexual experience. It is what the journey is leading toward, the most exciting moment of a wonderful trip, and the pinnacle of the sexual encounter. It's also the shortest phase of the sexual cycle, with rhythmic muscle contraction as its defining biological characteristic. At the moment of orgasm, muscles all over the body contract, although people tend to pay most attention to the spasms in very specific areas . . . .

A PEAK EXPERIENCE

In some ways the male and female bodies almost seem to be designed to work for the pleasure and satisfaction of their sexual partner. For example, semen is a potent neutral buffer, which is necessary to prevent sperm from dying in the acidic vaginal environment, but it is also an effective lubricant. The vaginal opening tightens involuntarily as intercourse progresses, which enhances pleasure for both partners, as well as encouraging male ejaculation. Men's and women's bodies are therefore not only optimized for individuality, but for interaction.

THE VAGINA PRODUCES LUBRICATION TO MAKE PENETRATION LESS DIFFICULT.

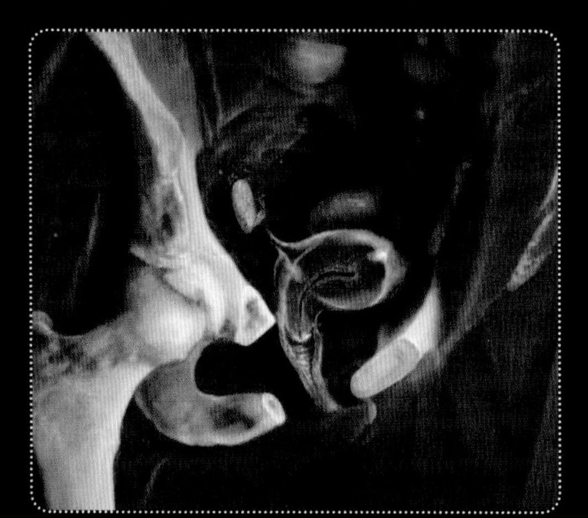

FAR FROM BEING A SMOOTH TUBE, THE VAGINA CONTAINS MANY FOLDS AND RIDGES.

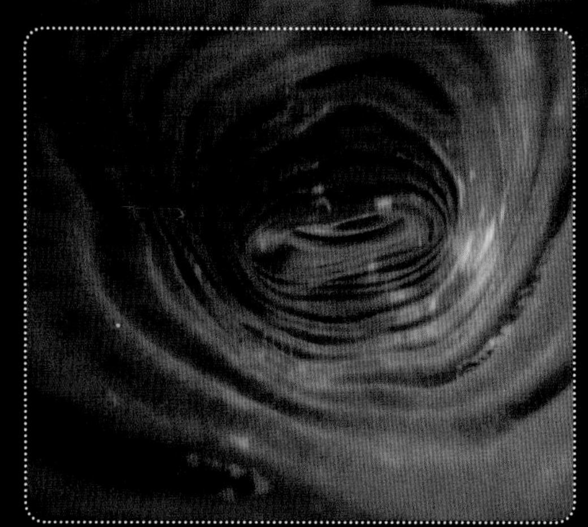

THE DIAMETER OF THE PENIS MAY EXPAND SIGNIFICANTLY WHEN IT GOES FROM FLACCID TO ERECT.

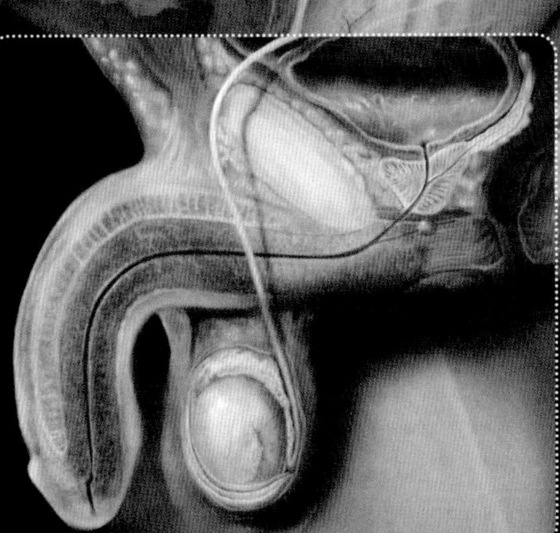

THE ERECT PENIS MAY POINT IN DIFFERENT DIRECTIONS AS A MAN AGES. IN YOUNGER MEN, IT MAY ACTUALLY POINT UP TOWARD THE STOMACH.

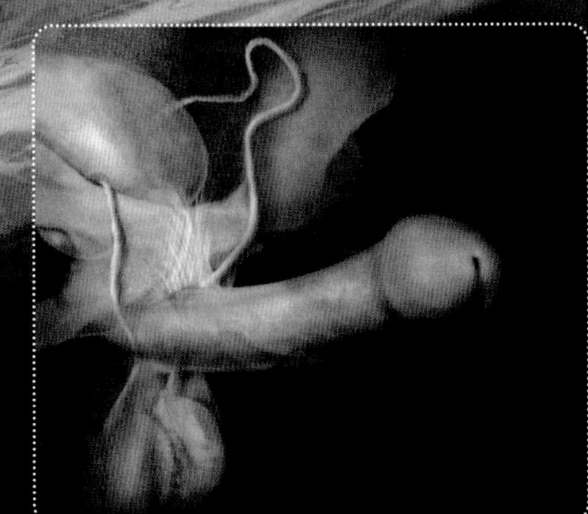

NERVES OF THE PENILE GLANS

The moment of orgasm is no time for ironic comments.
—MASON COOLEY,
City Aphorisms, Second Selection

IT'S MY PLEASURE

Orgasm may, in many ways, be a function of the brain and nervous system. The sexual organs contain a dense concentration of nerves, which contribute extensively to the sensations of passionate pleasure. In particular, the glans of the penis and its female counterpart, the head of the clitoris, are some of the most densely innervated parts of the human body. The interior of the vagina is sensitive as well, although much less so than the clitoris, or even the back of the hand. The most concentrated areas of vaginal sensation are at the top of the entrance, near the clitoris. The inner vagina is much less sensitive to touch, and many women don't enjoy contact with their cervix during intercourse. However, some research suggests that more than 25 percent of women may find cervical stimulation pleasurable.

In men, orgasm and ejaculation are closely, but not exclusively, linked. Male ejaculation actually occurs in two phases. The first phase takes place when the prostate, seminal vesicles, and vas deferens contract and force the ejaculate into a bulb at the base of the urethra. The second phase occurs when the urethral bulb and penis contract rhythmically to expel the ejaculate out the tip of the penis. Nonetheless, it is possible for some men to experience orgasm without ejaculation, and these men may be more likely to have multiple orgasms.

In women, the most profound change during the plateau phase is the formation of the orgasmic platform. In contrast to the earlier ballooning of the inner vagina, the orgasmic platform is actually formed by a swelling and tightening of the outer vagina. It is the orgasmic platform that has the most noticeable rhythmic contractions. A woman may experience three contractions, or a dozen, and the muscles of the anus and uterus may contract as well. Women are far more likely to be multi-orgasmic than men, and some women are actually capable of ejaculation during orgasm—but the precise nature of what and how they're ejaculating remains a subject of scientific debate.

NERVES OF THE GRAFENBERG SPOT ("G-SPOT")

NERVES OF THE CLITORIS

A SERIES OF MRIS OF TWO PEOPLE ENGAGED IN VAGINAL INTERCOURSE. YOU CAN SEE THE CHANGE IN POSITION OF THE PENIS AS IT PENETRATES THE VAGINA, AND AS THEIR BODIES MOVE CLOSER TOGETHER.

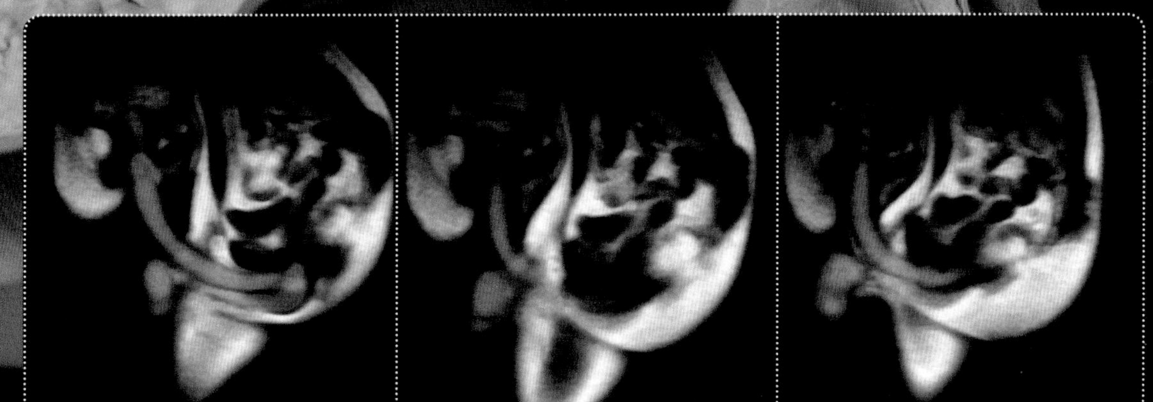

FLUID DYNAMICS

An important function of vaginal lubrication is to enhance the experience of sex. The vagina contains no secretory organs. How, then, does the vagina become lubricated? As the blood vessels in the vaginal walls dilate and increase blood flow, fluid seeps through into the vagina in a process known as transudation. Basically, increased blood flow causes the vessels to sweat.

Does female ejaculation exist? Many people think so. The pornography industry certainly does, and markets numerous videos to instruct women in the technique. The scientific community, on the other hand, is less certain. Major gynecologic journals scoff at the evidence, while sexual health textbooks accept it as fact.

Male ejaculate is a complex fluid designed to nourish and safeguard sperm during their journey through the female reproductive tract. The first component of the secretion is mucin, a potent basic buffer made by the Cowper's glands. This buffer is necessary because from the moment sperm enter the vagina, they are in danger. The normally acidic vaginal pH is fatal to them, and they can survive only because the buffering capacity of semen temporarily raises the pH to neutral so that sperm can make their way into the relatively less hostile environment of the uterus. Prostaglandin, one of the seminal components made in the prostate, actually causes uterine contraction— which may aid in the uptake of sperm. The prostate also contributes other supportive compounds, including immunoglobulins. Finally, the seminal vesicles make their contribution to the mix by producing the majority of the volume of the ejaculate. One of the most important components of

THE UTERUS FITS SNUGLY ON TOP OF THE BLADDER, AND THE URETHRA RUNS ANTERIOR TO THE VAGINA.

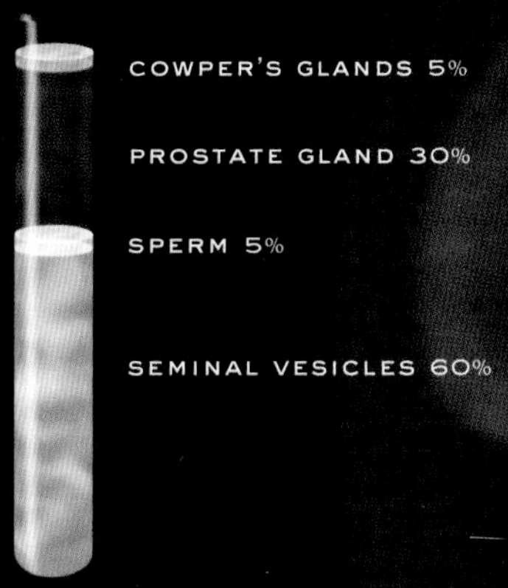

SEMEN IS A COMPLEX FLUID. COMPONENTS ARE COLLECTED FROM NUMEROUS GLANDS BEFORE IT IS EJACULATED FROM THE PENIS.

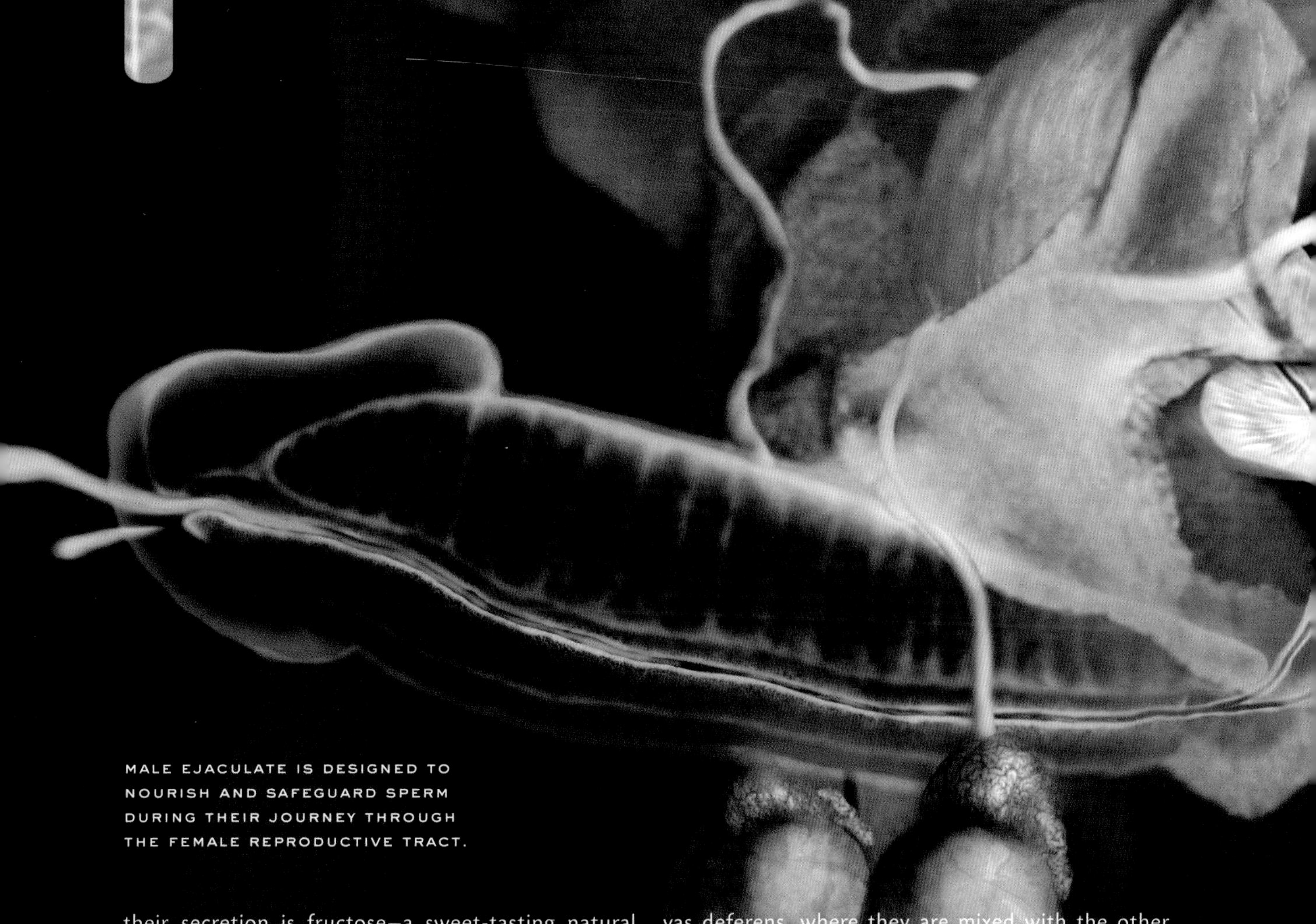

COWPER'S GLANDS 5%

PROSTATE GLAND 30%

SPERM 5%

SEMINAL VESICLES 60%

MALE EJACULATE IS DESIGNED TO NOURISH AND SAFEGUARD SPERM DURING THEIR JOURNEY THROUGH THE FEMALE REPRODUCTIVE TRACT.

their secretion is fructose—a sweet-tasting natural sugar. The reproductive voyage requires a great deal of energy, and this special sugar not only nourishes the sperm on their exhausting journey, but may also contribute to semen's unique taste.

The path that sperm travel during ejaculation is a long and winding road. Sperm take approximately three months to mature. However, despite traveling through the epididymis, they are still not ready to do their job. Before ejaculation, they move through the vas deferens, where they are mixed with the other components of semen in preparation for being expelled through the urethra. Actual ejaculation requires the coordination of a series of valves, which first allow the semen into the ejaculatory ducts—the emission phase—and then expel it. Even after ejaculation, the sperm still aren't prepared for fertilization. The process of sperm activation actually continues up until the moment they reach the egg.

RESOLUTION INTO REPRISE

Resolution is the end of the sexual journey, where the body returns to its unaroused state. The heartbeat starts to slow, breathing deepens, and muscles begin to relax. The first thing that women experience is the release of the pressure in their breasts and the return of their clitoris to its resting position. Then the orgasmic platform also relaxes and begins to shrink, as the vagina and uterus go back to their resting shapes.

Resolution takes, on average, between 15 and 20 minutes, but it can take longer in women who have not had an orgasm, because the explosive release of orgasm helps allow the physical tension to subside. For women who experience multiple orgasms, however, even the slightest bit of stimulation can move them back into active arousal, repeating the sexual cycle until they are exhausted and replete. Each orgasm makes it easier to experience the next as the body remains at an elevated level of sexual response.

For men, resolution is a two-stage process. During the first, although the stiffness of the penis releases as the corpora cavernosa empty, the penis maintains its aroused size. The erection doesn't fully subside until the second, much slower stage, when the corpus spongiosum and glans empty. During resolution, most men enter a refractory period where they cannot orgasm again. This period can range from several minutes to more than a day, but there are some men who do not experience a refractory period at all. For these men, loss of erection does not automatically follow orgasm, and they are capable of multiple orgasms within a short period of time. In multi-orgasmic men, ejaculation and orgasm may be independent, or the quantity of ejaculate may decrease with each successive orgasm.

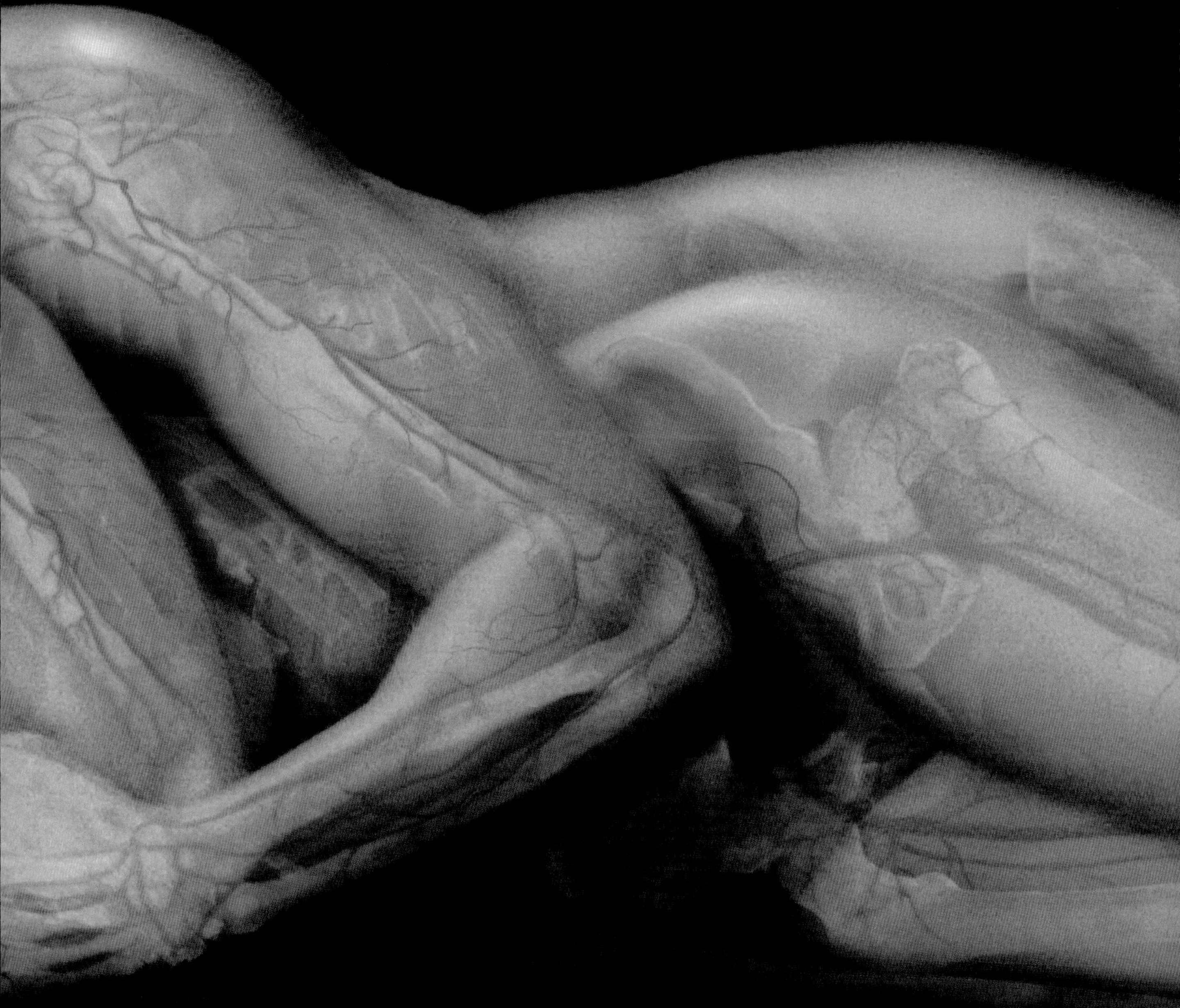

THE MAJOR NERVES OF
THE FEMALE PELVIS

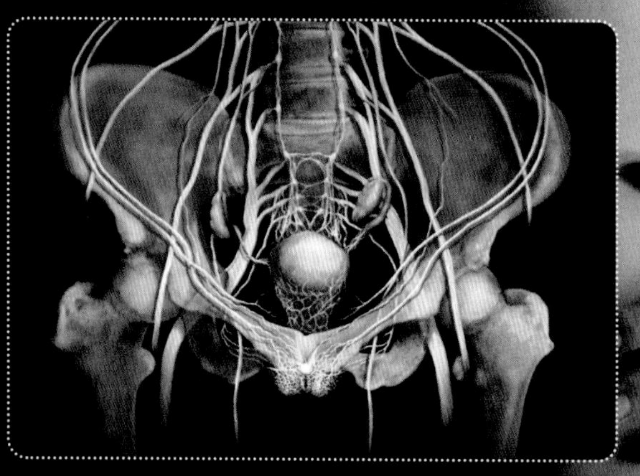

THE MAJOR NERVES OF
THE MALE PELVIS

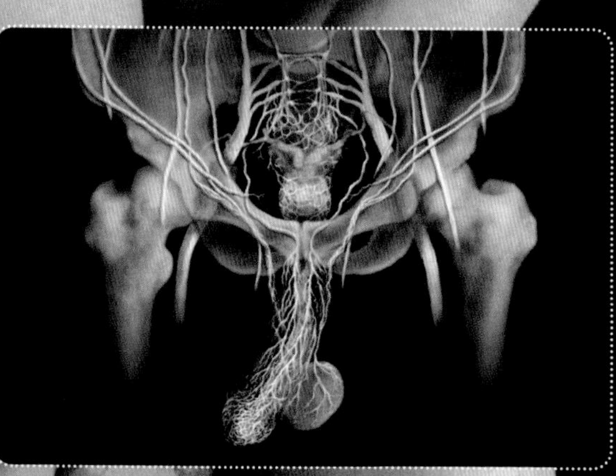

FEELING HORMONAL?

Sometimes the phrase "sexual decision making" seems like an oxymoron, because many people do not act rationally about sex. They let their desires take over the thinking portion of their brain, and then blame their mistakes on factors they claim are out of their control. Can your body set off cascades of hormones that make your libido surge, that make you want sex to the point of distraction, and that cause your genitals to ache with the desire for touch?

The feedback system involved in sexual excitement has many unique characteristics that reveal the importance of reproduction in human lives. Even the connections formed by sexually sensitive neurons of the brain are special. These neurons are linked by a unique blood vessel portal system so that they can communicate directly with the specialized cells of the pituitary gland and induce them to begin a hormone cascade.

Sexual arousal is both conscious and unconscious. Women have been shown to be able to think themselves to orgasm, but the actual process of arousal is controlled by unconscious cascades. Specifically, the body's drugs of choice when it comes to sexual arousal are androgens such as testosterone. Androgens are produced not only in the ovaries, but in the adrenal cortex, and they are what revs up a woman's libido (a Latin word meaning desire or lust). They do this in a number of ways: by increasing genital skin sensitivity, communicating to nerves in the spinal column and brain, converting to estrogen to modulate the vaginal environment, and regulating neurotransmitter release to affect sensation and mood.

In men, libido is separated hormonally from the ability to ejaculate and orgasm. This became clear hundreds of years ago when it was found that some eunuchs could still become sexually aroused, even though their lack of testicles made ejaculation impossible. Testicular testosterone is clearly extremely important for male sexual arousal, so how could they function without it? The answer is that androgens are also produced elsewhere in the body. The liver, the skin, and the adrenal glands all make androgens that are weaker than testosterone, but which still can have a sexual effect. While testosterone is the most effective stimulant of both sex drive and ability in men, it is important to consider that sexual activity in men may also be independent of hormones. Testosterone replacement therapy is an effective way to enhance the sexual lives of men who have decreased testosterone production due to aging or other health factors.

During sex, the brain shifts its attention to the sexual act. . .which seems much more interesting than worrying about cause and effect, actions and consequences. During the sexual act, people are too involved in their activity to be concerned about things that would bother them in the more sober aspects of their daily lives. Worrying about where someone's mouth has been doesn't seem terribly important when that mouth is on you.

Injections of testosterone increase women's arousal and vaginal responsiveness to hardcore pornographic films.[4]

TOUCH ME

Although it's hard not to laugh at the Emo Phillips line, "I used to think that the brain was the most important organ in my body, but then I realized who was telling me this." The brain actually *is* the most important organ when it comes to sex. Not only is it the home of thought and fantasy, but it is also where all the signals from the rest of the body are processed and perceived as pleasure or pain. It affects the production of sex hormones, determines who and what you find attractive, and helps you sense arousal in others. Its complexity is what takes sex from reproduction to remarkable.

A hand may stroke your calf, or lips may touch your neck, but all touch is perceived in the same place—in the brain. Whether your skin is stimulated by a breath or a finger, nerves in that location fire, signals are transmitted through the cranial nerves to your brain, and your brain recognizes that you have been touched. During sexual arousal, the brain becomes even more responsive to touch and other sensations, answering each stimulus by increasing blood flow to the most sensitive areas, and setting up a feedback spiral that allows sexual arousal to ascend to its peak.

Touch sensors, types of sensory nerves, are specialized to sense pressure, deep and light touch, vibration, temperature, or stretch. Concentrations of each type of sensory nerve vary in different areas of the skin. Any or all of these sensations can occur during a sexual interlude. An ice cube trailed across the skin, the weight of a lover's arm—the body receives every type of sensation through its own unique receptor, which then sends chemical and electrical signals to relate its encounter to the brain. The lips, in particular, contain a high density of nerves, and a large portion of the sensory apparatus of the brain is devoted to analyzing their tactile experiences. This is one of many reasons why the mouth plays such an important role in the sexual experience.

Food both sustains and satiates. Its purpose is to nourish, but the textures and tastes upon the tongue also can stimulate feelings of pleasure. Like smells, tastes stimulate the brain's pleasure centers (the limbic system), and emotional reactions can become associated with certain textures or flavors. Kissing provides your brain with sensations of both touch and taste, and it can be an extremely erotic activity. In many cultures, the image of a kiss is synonymous with not just passion, but with true romance and love. It's not a universal behavior, however: some cultures actually consider kissing a disgusting practice—unhygienic, repulsive, even suggestive of cannibalism!

DIFFERENT TYPES OF SENSORY NERVES HAVE DIFFERENT FUNCTIONS.

PACINIAN CORPUSCLE—LIGHT TOUCH

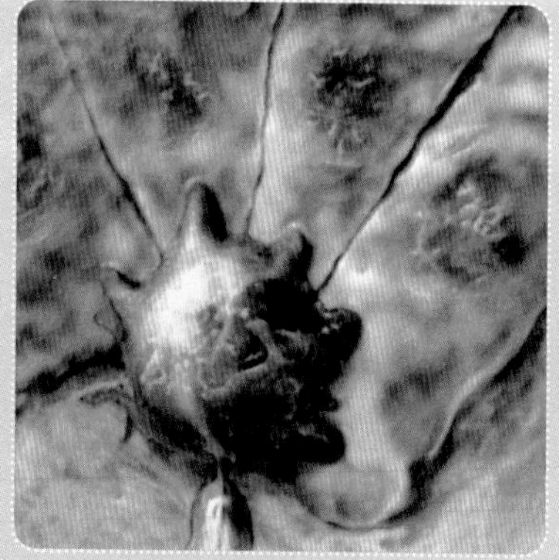

MERKEL'S DISC—DEEP TOUCH AND PRESSURE

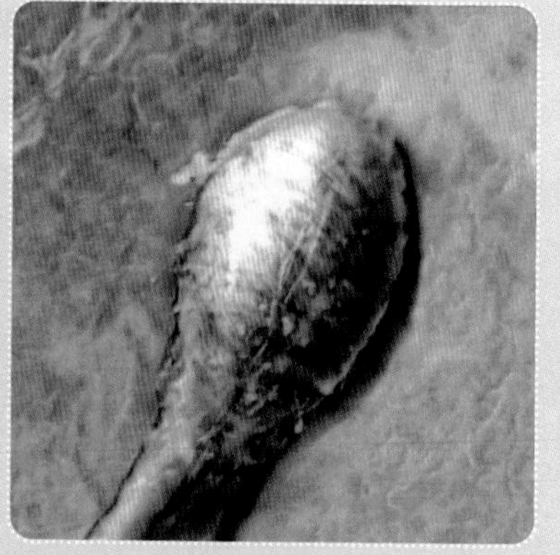

MEISSNER'S CORPUSCLE—LIGHT TOUCH

Then I did the simplest thing in the world. I leaned down . . . and kissed him. And the world cracked open.
—AGNES DE MILLE

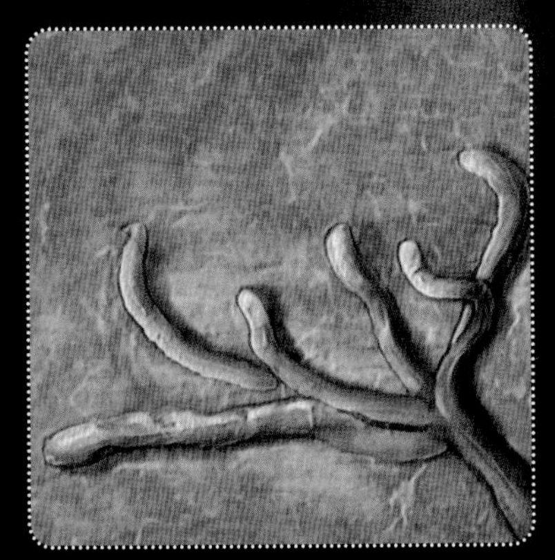

FREE NERVE ENDINGS—TOUCH, PRESSURE, PAIN, TEMPERATURE

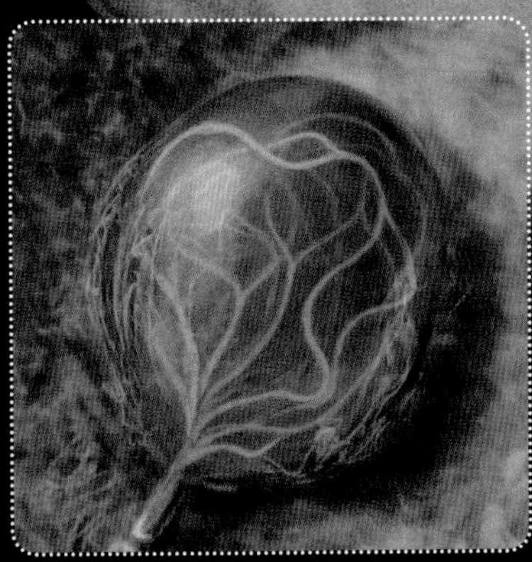

KRAUSE'S END BULB—LIGHT TOUCH AND COLD

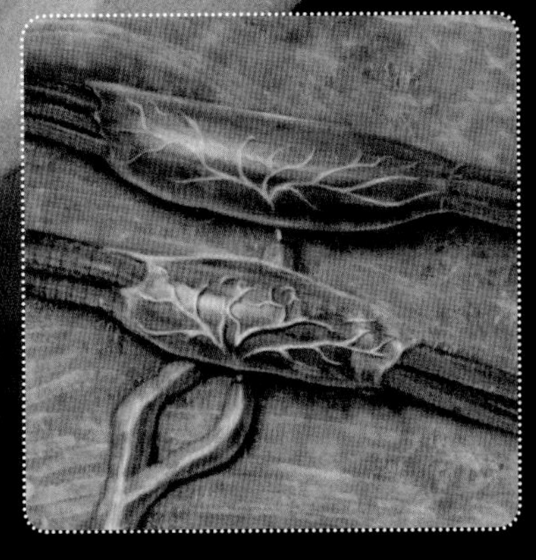

RUFFINI ENDINGS—DEEP TOUCH, PRESSURE AND HEAT

For most people, sight is a very important component of sexual arousal. The things that "turn them on" may vary, but there's almost always some type of visual that can spark a sexual response. It's not just a stereotype, however, that men tend to be far more interested in sexual imagery than women; it's a scientific fact. Even though looking at a suggestive picture activates the same pathways in men's and women's brains, men tend to have higher levels of electrical response even when women report being more physically aroused. This suggests that something that occurs during development affects how men and women respond to visual stimuli. While they may both find an image sexually arousing, they respond in biologically different ways.

What, and how, images are processed into arousing stimuli remains poorly understood. Light is received by the rods and cones of the retina and transmitted to the visual cortex for interpretation, but how these images stimulate sexual thought is still

THE LOOK OF LOVE

VISUAL CORTEX

OPTIC NERVE

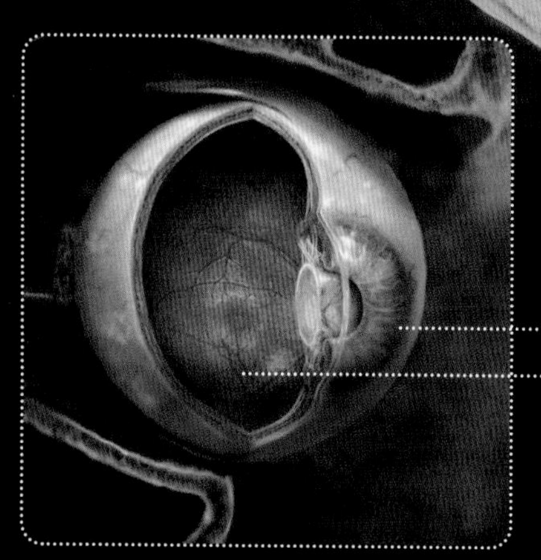

LIGHT ENTERS THROUGH THE IRIS AND IS TRANSLATED BY LIGHT SENSORS THAT RESIDE IN THE BACK OF THE RETINA INTO NERVE IMPULSES. THE IMPULSES TRAVEL THROUGH THE OPTIC NERVE TO THE VISUAL CORTEX OF THE BRAIN.

a matter of debate. The visuals that people find sexually stimulating vary between sexes, cultures, and even age groups, and seem to be learned from the environment rather than being instinctive. In men, there seems to be a direct link between seeing a visually arousing image and activation of the testosterone-responsive hypothalamus, but in women the connection between sight and arousal is less well understood. But it's clear that visual stimulation can be erotic to both sexes, even if no one can explain how.

Why does the sound of a quickened breath or a throaty whisper send shivers down a person's spine? No one knows, but it's very certain that something stimulating is going on. Even more interestingly, several studies have found that the way people process sound varies not only by sex but also by sexual orientation, suggesting another link between sensuality and aural pleasure.

COCHLEA

ACOUSTIC NERVE

AS SHE WHISPERS INTO HER LOVER'S EAR, IT ACTIVATES SPECIALIZED CELLS IN THE COCHLEA THAT SENSE THE MOVEMENT OF HER BREATH AND TRANSMIT HER WORDS TO HIS BRAIN VIA THE ACOUSTIC NERVE.

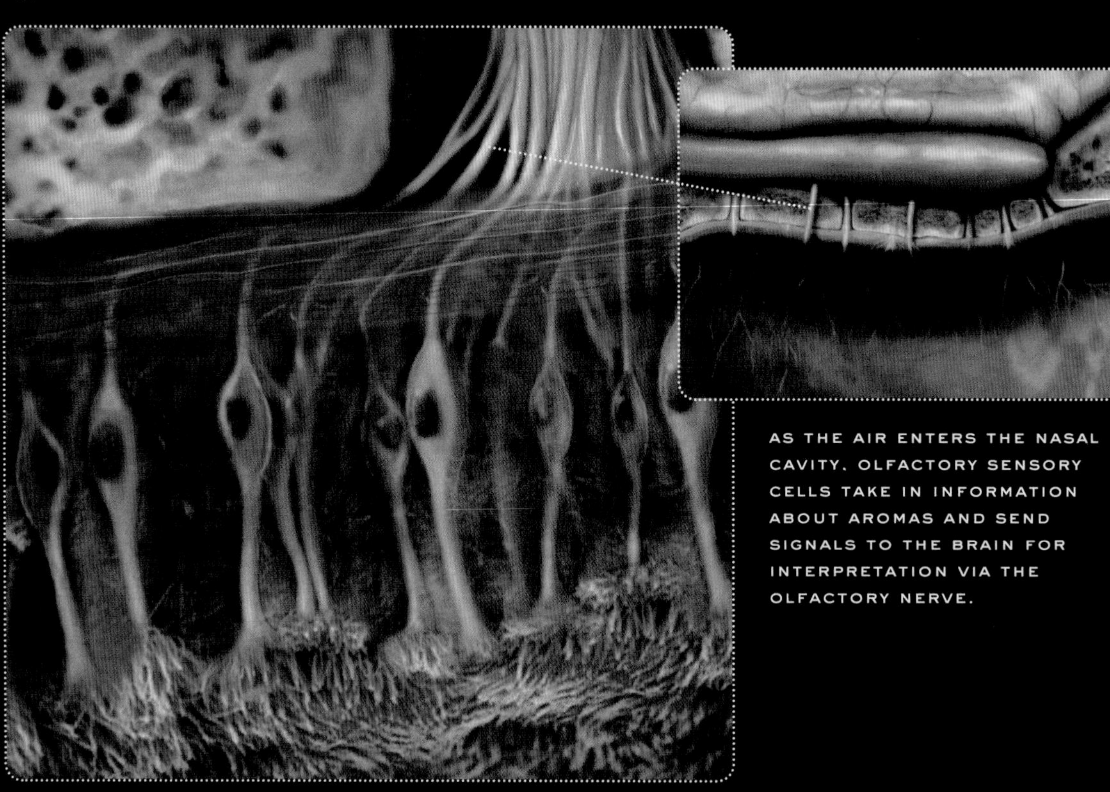

AS THE AIR ENTERS THE NASAL CAVITY, OLFACTORY SENSORY CELLS TAKE IN INFORMATION ABOUT AROMAS AND SEND SIGNALS TO THE BRAIN FOR INTERPRETATION VIA THE OLFACTORY NERVE.

THE SCENT OF A (WO)MAN

Smell is one of the most potent senses for stimulating the human brain. Odors are profound triggers of emotional and sense memory, and they have both conscious and subliminal effects on the mind. Our emotional response to odors is far stronger than to their related visual or auditory stimuli.

Some of the most interesting research on smell has involved pheromones and how they affect human behavior. Pheromones are chemical compounds detected by the nasal sensory neurons within the nose, although pheromones are usually odorless. They are defined by the effect they have on social and sexual behavior. Pheromones have been shown to affect the timing and synchronization of the menstrual cycle in groups of women, mother-infant bonding, and gender-based sexual repulsion and attractiveness. Interestingly, research suggests that the brains of homosexual men respond to male pheromones in a manner similar to that of heterosexual women, but not to female pheromones the way that heterosexual men do.

In contrast to pheromones, odors are processed by the olfactory neurons and the olfactory cortex. They are then interpreted both consciously through the neocortex, and emotionally through the limbic system. It is through the limbic system that odors stimulate memory, attraction, and gut reaction, because of previous associations between that odor and emotional states or activities. Pheromones, and other sexually attractant odors, appear to work at

PHEROMONE PRODUCTION

THE TONGUE, BECAUSE IT SENSES TASTE, PRESSURE, AND HEAT, CONTRIBUTES MUCH TO THE ENJOYMENT OF KISSING.

least in part by stimulating the hypothalamus which, with its intimate connection to hormonal control, is a major sexual component of the brain.

The odors of sex are particular stimulants of memory and arousal. Studies have shown that a woman's body odor is most attractive to men when she's in the fertile stage of her menstrual cycle. Evolution has distilled a stimulus that encourages sex when it will be most productive. There is also a reason why women so enjoy snuggling into the center of their partners' chest. One of the apocrine glands is located there, and the scent produced by these glands is regulated by the sexual hormones. In other words, the pheromones created there at any given moment are stimulated by, and in turn enhance, the current mood. It is a comfortable place to

All the senses are involved in sexual desire, and sexual desire is born from both biological and cultural stimulation. What we are attracted to, what turns us on, the things that bring us passion are affected as much by personal history as by biology. Erogenous zones, for instance, are defined as those places on the body that, when touched, give rise to sexual desire. You may think that those locations are mapped and certain, but a person from Asia will give you very different descriptions of their erogenous zones than a friend from North America. We are conditioned by our environment to believe that certain stimuli and behaviors are sexual, and research has shown that individuals can be trained to be aroused by smells, sensations, and images that have no direct relationship to their

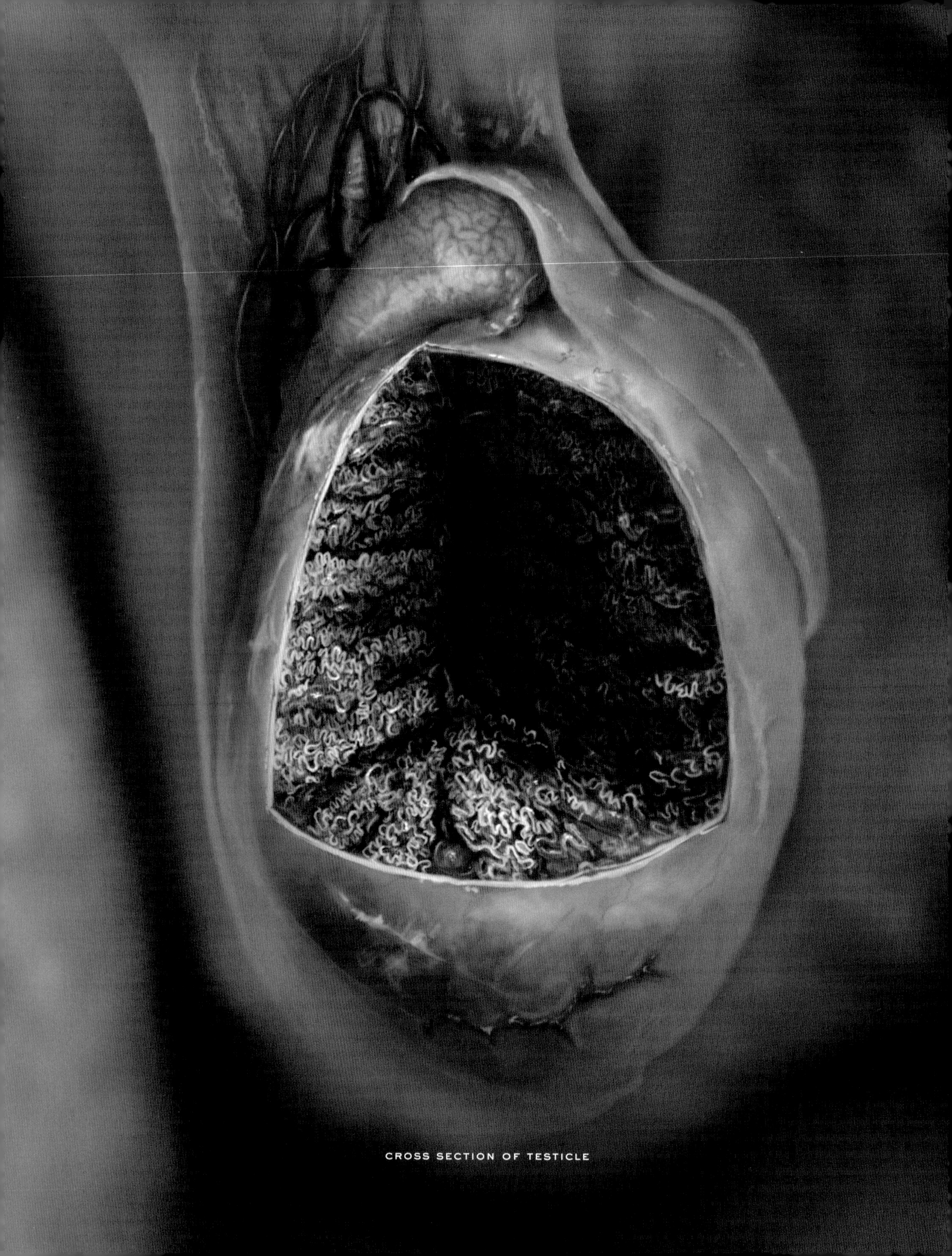

CROSS SECTION OF TESTICLE

CROSS SECTION OF OVARY

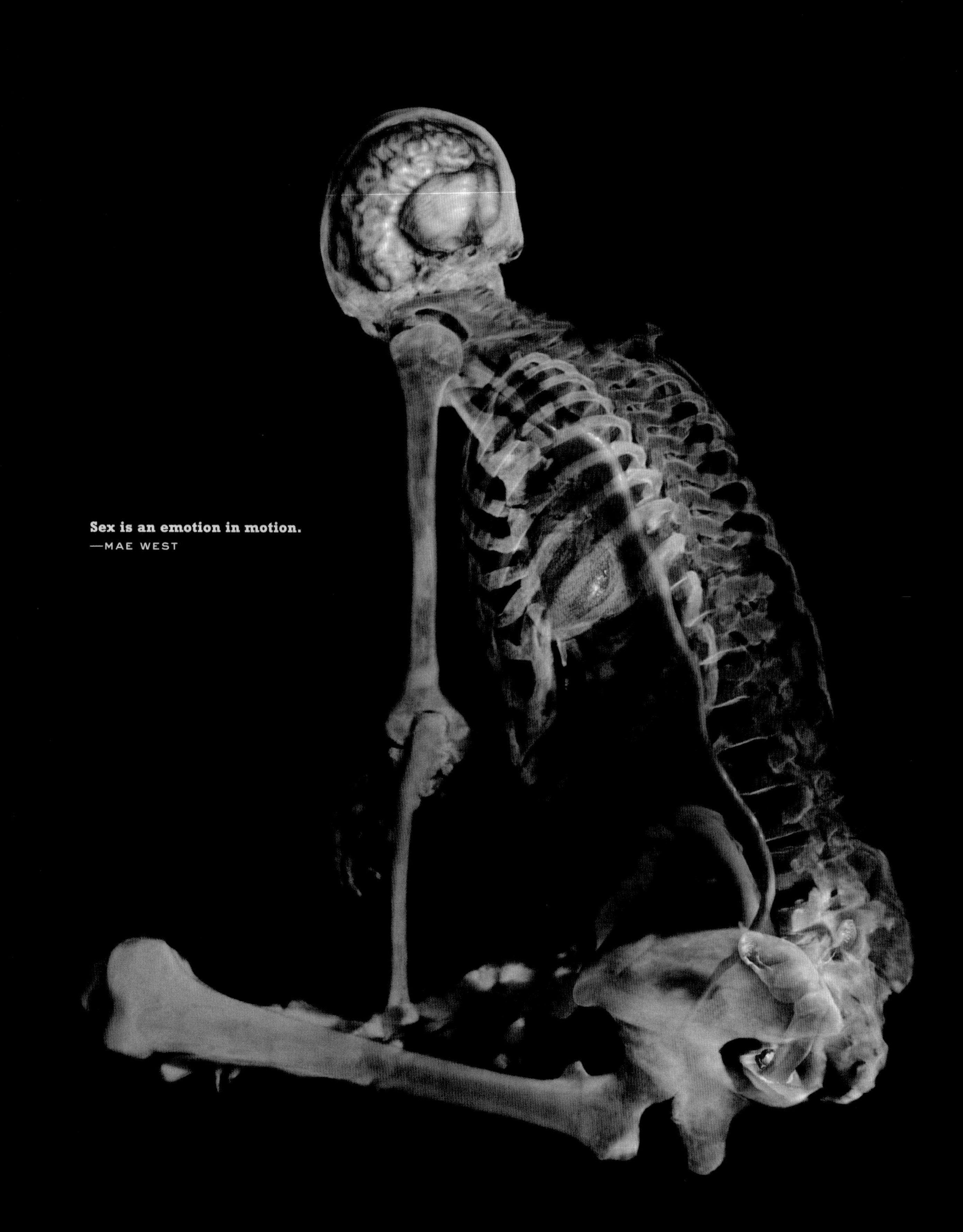

Sex is an emotion in motion.
—MAE WEST

3

IT'S OFTEN THE MIND THAT MATTERS

There are times when you're just not interested in sex. You know it would be fun once you got into it, but on a bad day the passion can seem much too far away. Sex starts in the brain before it moves into the body, and there comes a point where a healthy sex life isn't possible without a healthy life. Sex ceases to be enjoyable, and instead becomes work.

LET'S HAVE A CONVERSATION

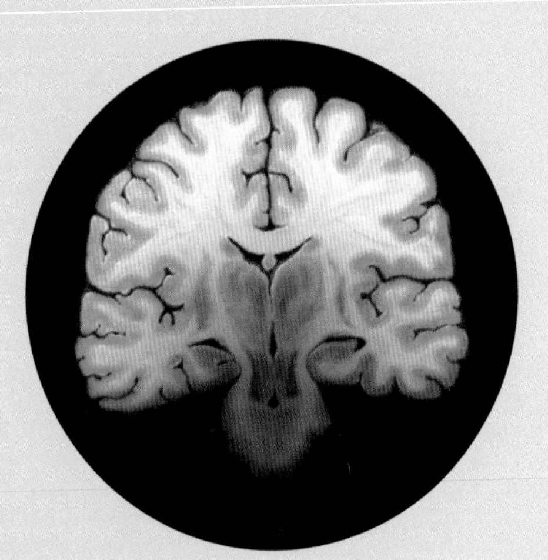

IT'S ALL IN YOUR HEAD: The brain is, in many ways, the most important sexual organ. Not only does it control the realm of fantasy, but it fuels and maintains the sexual machinery as well.

I HAVE A HEADACHE: When life is stressful or depressing, a sexual interlude might cheer you up . . . if only you could get in the mood.

IT'S NOT JUST WINE THAT IMPROVES WITH AGE: Sexual function changes with age, but sometimes for the better. It's not all downhill from here.

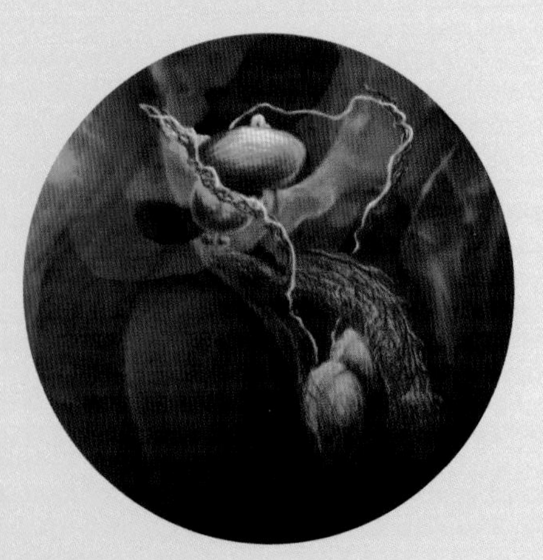

BLAME IT ON YOUR NERVES: Sex may start in the brain, but then the entire nervous system gets into the act.

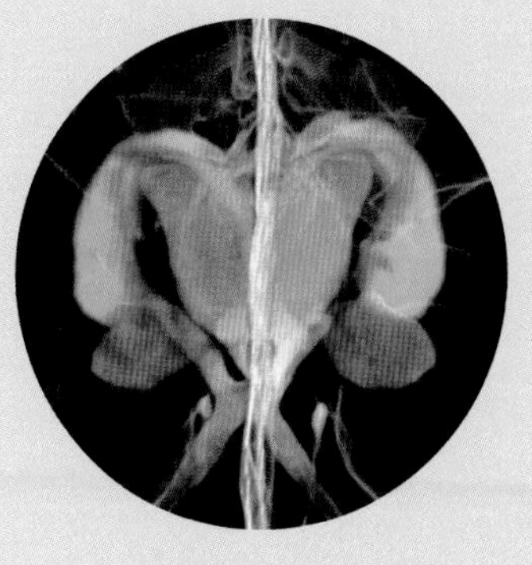

IT SEEMED LIKE A GOOD IDEA AT THE TIME: Sex, drugs, and rock and roll don't go together as well as you might think. Alcohol or other drugs may get you in the mood, only to take away your motivation—and your good judgment, too.

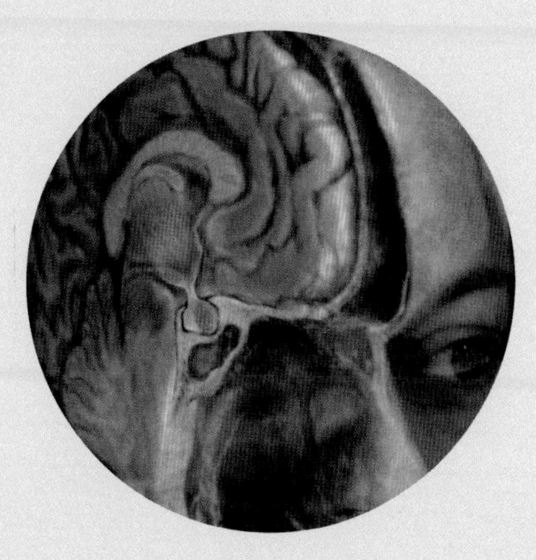

SEXUAL CHEMISTRY: It's clear what turns you on, but how exactly does that get you sexually aroused? It's amazing how a frothy little concoction of chemical impulses can be so potent.

IT'S MORE THAN JUST A BROKEN HEART: Not taking care of your health can take away your sex life. Failure to control chronic diseases such as high blood pressure and diabetes can deprive you of sexual satisfaction.

Most people think of the nervous system as an electrical highway of sorts that conducts information about physical experience from different parts of the body to the brain. In this model, the central nervous system, or CNS, is the central processing unit of the body. Consisting of the brain and spine, the CNS connects to the rest of the body via the peripheral nervous system (PNS), and information and instructions flow back and forth in both directions.

The role of the nervous system in sexual arousal is both conscious and unconscious. Although much of arousal begins in the brain and is communicated to the genital area, there is also an element of reflex involved. Research on individuals with spinal cord injuries has demonstrated that even when there is loss of arousal controlled by the CNS, arousal by local stimulation can still remain intact. This is true even if the person cannot feel the sensation of being touched, as long as the relevant nerve pathways are still present.

Although sexuality has a reflexive component, we can override it with the activity of our brain. Thinking too much can stop the arousal process in its tracks just as efficiently as biological problems can. There is so much in a sexual encounter that can go wrong—thinking the wrong thoughts, not having enough blood flow, a failure of communication in the nervous system, a failure of communication with your partner. . . .If you think about it, it's pretty amazing that anyone ever manages to have sex at all.

THE SENSUAL PROCESSING UNIT

Sex is more exciting on the screen and between the pages than between the sheets.
—ANDY WARHOL, *From A to B and Back Again*

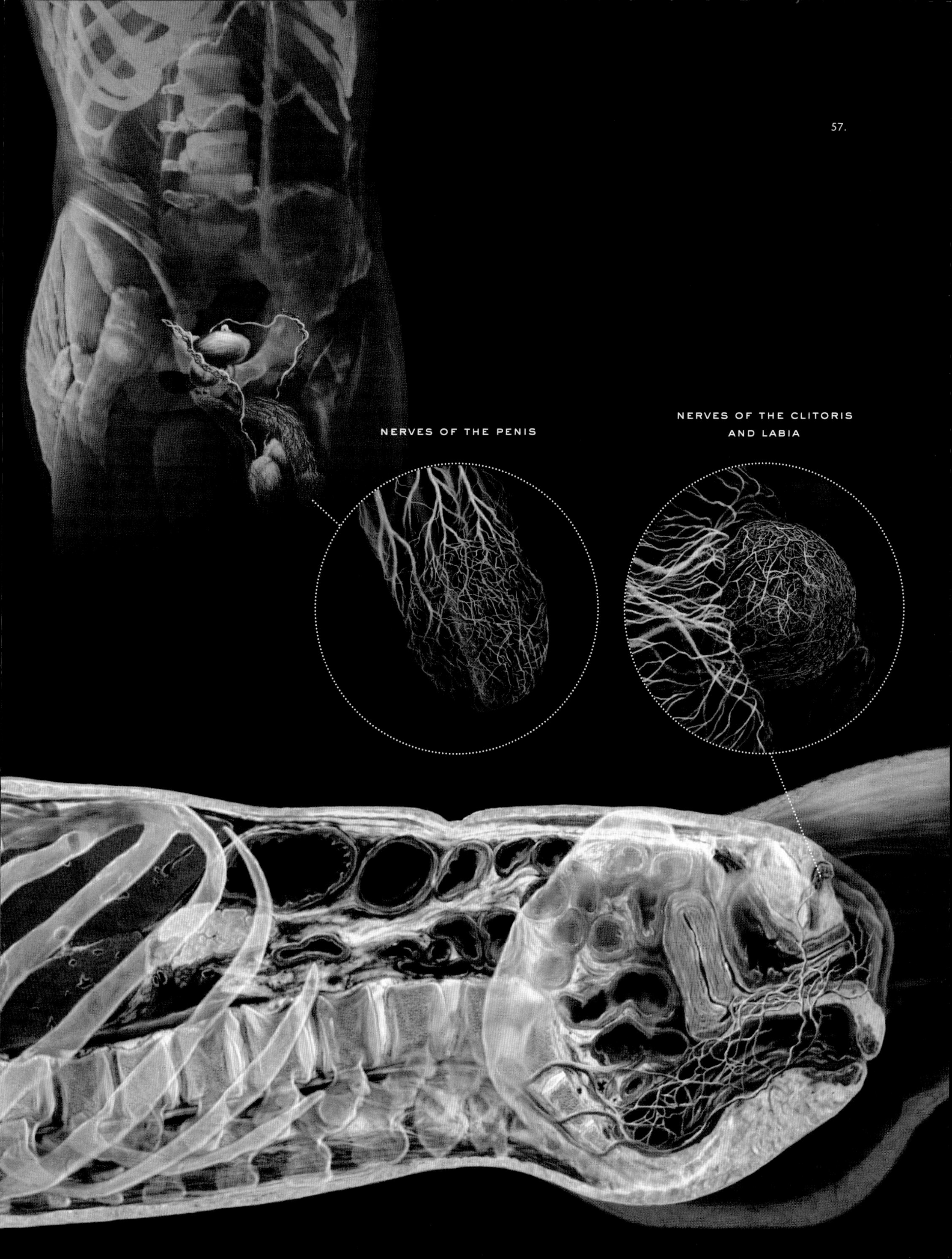
NERVES OF THE PENIS
NERVES OF THE CLITORIS AND LABIA

CROSS SECTION OF THE BRAIN

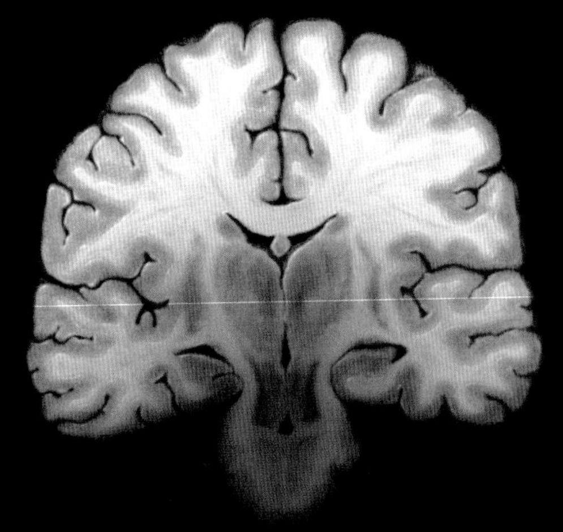

SEX AND STRESS DON'T MIX

Stress has a profound effect on sexual health. Not just a mental issue, stress has a strong biological component that can be measured by its effects on body levels of cortisol, a steroid that has various metabolic and anti-inflammatory effects in the body, and other hormones. By reducing circulating levels of sex hormones, stress can reduce sexual desire and function in men and women. Stress also has negative effects on the cardiovascular system. It increases blood pressure and causes vessels to contract, both of which further impair sexual function—remember that vasodilation, or dilation of vessels, is crucial to sexual arousal. Fortunately, reducing stress relieves its effects and allows people to resume their normal sexual lives.

Depression and other mental disorders are intimately linked with sexual function. First, the very parts of the brain that are involved in sexual arousal and function are those that are impaired in many individuals with mental illness. Second, many of the drugs used to treat the common mental illnesses of depression and schizophrenia also disable the pathways involved in sexual response. Depending on their precise chemical action, antidepressants have been shown to affect sexual interest, the ability to become aroused, the ability to reach orgasm, and the experience of orgasm. Fortunately, this problem can often be solved by switching to a different antidepressant. Anti-psychotic treatments for schizophrenia actually work to limit dopamine secretion by the limbic system. Dopamine is a neurotransmitter associated with mood enhancement, including sexual pleasure. Limiting dopamine can have the effect of damping down sexual feeling as well as other emotions.

Emotions themselves are, in other circumstances, the source of sexual difficulties. Sexual desire and sexual pain disorders in both men and women can have both psychological and organic origins. The lack of sexual interest that one woman has because of a low testosterone level may, in another woman, be the result of a traumatic sexual experience. Similarly, in men, premature ejaculation and related disorders are equally likely to be caused by anxiety or biology. Thought and worry can trap people in an unending cycle of sexual disorder: being afraid of, or uncomfortable about, sex makes it difficult to become aroused, which makes sex less enjoyable, which causes even more fear—and the more concerned you are about having trouble getting aroused, the more you think about it and analyze it, the harder it is to relax enough to let your body turn itself on. It's the centipede's dilemma: if he had to think about how to coordinate all 100 feet, he'd never manage to actually walk anywhere.

The mind and body are fundamentally linked in both the joys of sex and in its disorders, and it takes patience and sometimes professional help to break the cycle once problems become entrenched.

INFORMATION FLOWS FROM ONE NEURON TO ANOTHER ACROSS A SYNAPSE.

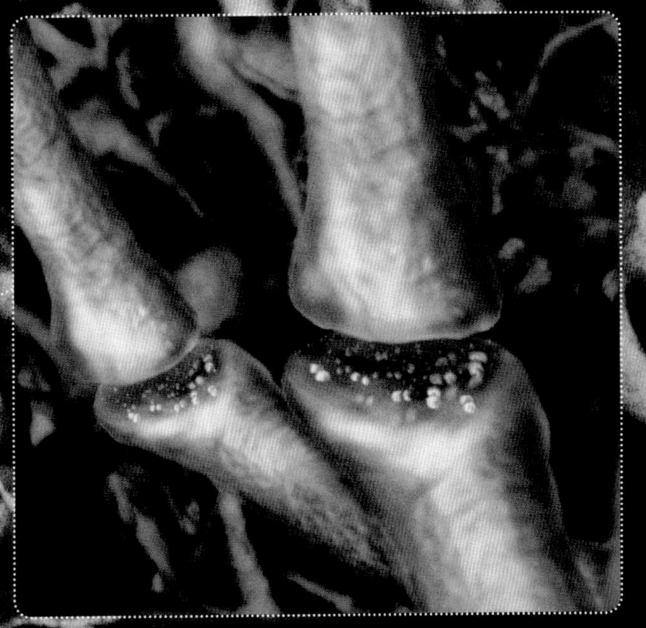

CLUSTERS OF NEURONS ARE PRESENT THROUGHOUT THE BRAIN TO RECEIVE AND PROCESS DIFFERENT SIGNALS.

OLDER AND WISER

As men and women age, their bodies change, and so do the ways they respond sexually. They may feel less attractive, or that they're not supposed to be as interested in sex, but sexual activity is as healthy a part of life in a person's 80s as it is in their 20s. Sex is as normal and natural for older people as it is for younger people. Having a healthy sex life can add richness and vigor to people's later years.

The sexual response changes with age, due in part to lower levels of hormones. In women, the excitement phase takes far more time. Lubrication is delayed, but although the amount of vasocongestion is lower than in younger women, it is usually sufficient for normal sexual function. Because of the lowered hormone levels, however, vaginal tissues are thinner, and the vagina is also less able to expand, which can lead to pain during penetration. The plateau phase is similar to that of younger women, although the uterus does not become as elevated or the orgasmic platform as pronounced. Fortunately, most postmenopausal women don't experience a change in their frequency of orgasm, although the number of contractions may be reduced, and resolution occurs more quickly afterwards due to the overall decrease in vasocongestion. Many of the negative effects of aging on female sexuality can be minimized by hormone therapy. However, for many women hormone therapy isn't appropriate, or even necessary. See Chapter Five for a discussion of hormone replacement therapy (HRT).

In men, aging primarily affects the duration and intensity of their sexual response. Erection takes longer, and may require more direct stimulation than when a man was younger. During the plateau phase, men experience less muscle tension, and the testicles do not elevate as close to the perineum (in men, the region between the scrotum and the anus). One benefit of aging is that men can, in general, sustain an erection for a longer period of time. They also have greater ejaculatory control, which may increase their and their partner's pleasure. Orgasm remains pleasurable, but men may have reduced numbers of contractions and less ejaculatory force and volume. Finally, the resolution phase is much faster in older men, and the refractory period may lengthen from a couple of minutes to several hours or even days.

For both sexes, however, the most important factor in maintaining a healthy sex life into old age is continuing to have sex. A regular sex life, on top of good general health, helps to keep everything working the way that it should.

In a 2001 study, 8 of 12 pre-menopausal women reporting decreased libido were found to have low or negligible testosterone levels. Oral administration of DHEA increased libido in seven of the eight. [5]

AS WOMEN AGE, THEIR SEXUAL RESPONSE CHANGES. LUBRICATION IS DELAYED, ESTROGEN LEVELS DECREASE, VAGINAL TISSUE THINS, AND THE VAGINA IS LESS ABLE TO EXPAND.

ESTROGEN MOLECULE

OVARIES

FALLOPIAN TUBES

UTERUS

VAGINAL WALL

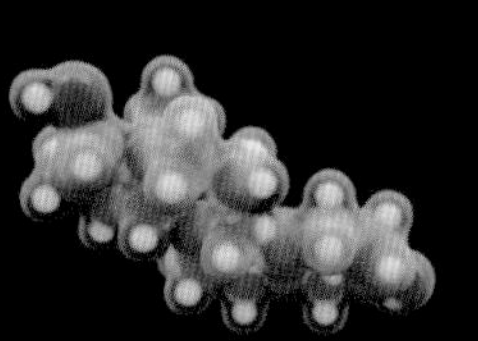

TESTOSTERONE MOLECULE

SPERM SWIMMING THROUGH A WOMAN'S FALLOPIAN TUBE

MEN ALSO EXPERIENCE CHANGES AS A RESULT OF AGING. HORMONE LEVELS MAY DROP, AFFECTING SEX DRIVE, SPERM PRODUCTION, AND OVERALL FEELINGS OF WELL-BEING.

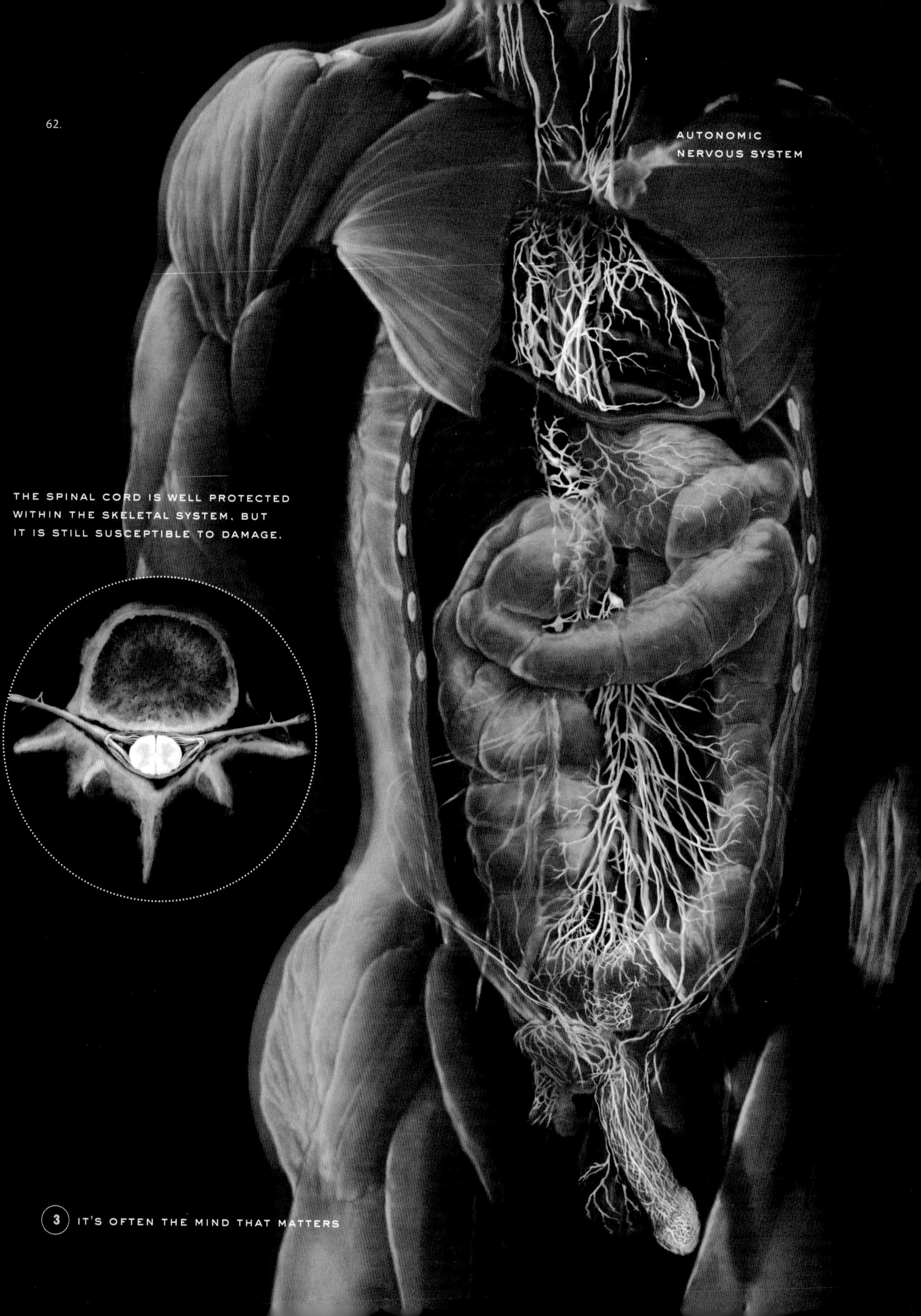

AUTONOMIC NERVOUS SYSTEM

THE SPINAL CORD IS WELL PROTECTED WITHIN THE SKELETAL SYSTEM, BUT IT IS STILL SUSCEPTIBLE TO DAMAGE.

YOU'VE GOT TO BE CALM TO GET EXCITED

The nervous system is not simply composed of the brain and the spinal cord. Although these are the central core of the system, the peripheral nervous system is where much of the information processed there comes from, relaying signals from the rest of the body. Three specific groups of peripheral nerves are particularly important in the sexual response: the pudendal (external genital), the hypogastric, and the pelvic. The pelvic nerves in women are the main vaginal and cervical sensory nerves, and in addition they control engorgement of the clitoris. In men, the pelvic nerves control erection of the penis. The hypogastric nerves are sympathetic nerves. Because they innervate the testicles, cervix, and uterine area, they are in charge of ejaculation and vaginal lubrication. Finally, the pudendal nerves control conscious contraction of the vaginal muscles and the muscles at the base of the penis, as well as the urethral and anal sphincters. The pudendal nerves also receive sensation from the penile skin, scrotum, labia, and clitoris.

Receiving and relaying sensation is not the only function of the peripheral nervous system. A very important component of the PNS, the autonomic nervous system, has another role entirely. It manages the systems of the body that need to function without conscious control.

The autonomic nervous system consists of two parts—the sympathetic and parasympathetic systems. Although both systems affect organs throughout the body, they tend to work in opposite ways. The sympathetic nervous system is responsible for what's known as the "fight or flight" reflex. Among other effects, arousal of the sympathetic system increases heart rate, breathing, and blood pressure as well as releasing the hormones commonly called adrenaline. In contrast, the parasympathetic system tends to inhibit the reflexive alarm responses of the sympathetic system.

Most research suggests that it is the parasympathetic, or calming, part of the autonomic system that is most responsible for the subconscious aspects of sexual arousal. However, some controversial research suggests that activation of the sympathetic nervous system by exercise or other stimuli may leave an individual more biologically susceptible to sexual arousal, even if the person doesn't consciously notice a difference. Translation: being scared may stimulate your interest in having sex, but you need to be relaxed to become physically aroused. It makes sense: if you were in a "fight or flight" situation, it wouldn't be very useful to your survival to become distracted by thoughts of sex!

NERVES OF THE FEMALE GENITALIA

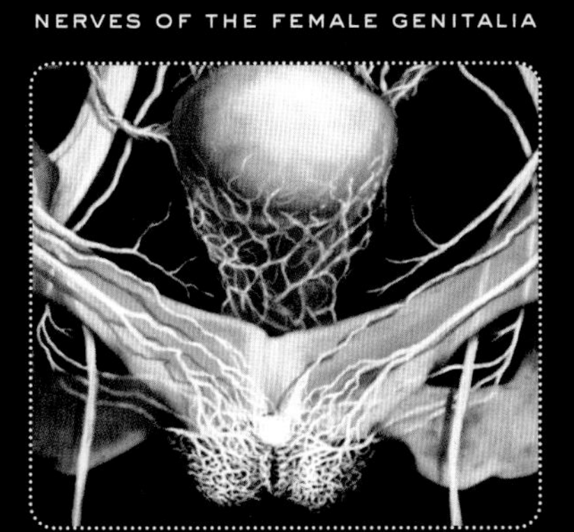

NERVES OF THE MALE GENITALIA

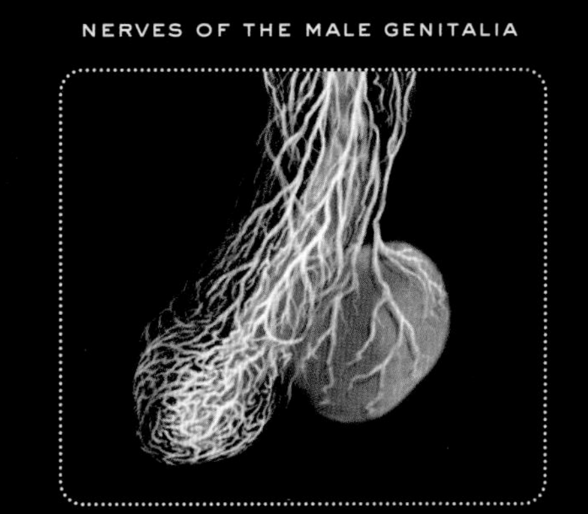

People believe that alcohol increases sexual function. Getting slightly drunk does, in fact, lower inhibitions and increase sexual risk-taking, but at higher levels alcohol actually decreases sexual ability. Alcohol is a systemic depressant that reduces both the capacity for arousal and the enjoyment of orgasm. It can also make it difficult for a man to achieve an erection. While alcohol may make it psychologically easier for a person to have sex, it makes it physiologically far more difficult—and it blunts good judgment, besides.

In the short term, drinking too much alcohol can make you do some pretty stupid things. But although most people think regrets and hangovers are the only lasting effects of alcohol use, in the long term excessive drinking can damage not only your body but your brain. People are frequently warned about liver damage from alcohol abuse, but very few individuals are ever told about the frightening fact that chronic drinking can actually cause loss of tissue throughout the brain.

Other legal intoxicants also affect a person's sexual function. A groundbreaking study on the causes of sexual dysfunction found that for any given age of men, the most important risk factor for erectile difficulty was cigarette smoking. Contrary to the image of the cigarette as an accessory to the sexual act—a signal for flirtation, or a way to ignite a literal afterglow, smoking is actually harmful to your sexual health. Smoking is a major risk factor for impotence in men, affecting the health of the vasculature, overall blood flow, and male hormone levels. Smoking

A FEW DRINKS AND OTHER BAD IDEAS

REGIONS OF THE BRAIN AFFECTED BY ALCOHOL USE

ALCOHOL MAY MAKE THE DECISION TO HAVE SEX SEEM EASIER. BUT IT MAKES THE ACTIVITY MUCH MORE DIFFICULT.

FRONTAL LOBE

HIPPOCAMPUS

PONS

CEREBELLUM

MEDULLA OBLONGATA

When you're drunk, you'll have sex with someone you wouldn't have lunch with.
—KOREN ZAILCKAS

also has negative effects on female sexual function. By reducing estrogen levels, it increases the risk of vaginal dryness, painful periods, and hormonal contraceptive failure. What's more, smoking decreases fertility in both men and women. Although it may seem erotic to light a woman's cigarette with a match struck to flame, in the long run the sexy effect is overshadowed by the very real loss of your health.

In the 1950s and 1960s, methamphetamines (popularly known today as crystal meth) were used as a treatment for obesity and depression, but today the drug is causing more than its fair share of physical and mental health problems. In the beginning of 2006, the United Nations declared that use of crystal meth was approaching pandemic proportions—with frightening health consequences, both direct and indirect. Methamphetamine use is dangerous, and users risk numerous health issues, including stroke, cardiac arrhythmia, paranoia, hallucinations, and structural changes to the brain. It's also the leading cause of drug-related emergency room visits in several states. Why do people use methamphetamines? Methamphetamines are easy to make from commonly available ingredients—labs can avoid the very real possibility of explosions. Methamphetamines make you feel good, keep you awake for days, and turn you on—which leads to their indirect consequences. Users want sex, and they tend not to be too careful about how they get it. Crystal meth use is associated with a significantly increased risk of STD transmission, because people taking it are far too intent on having sex to think about using condoms and practicing safer sex.

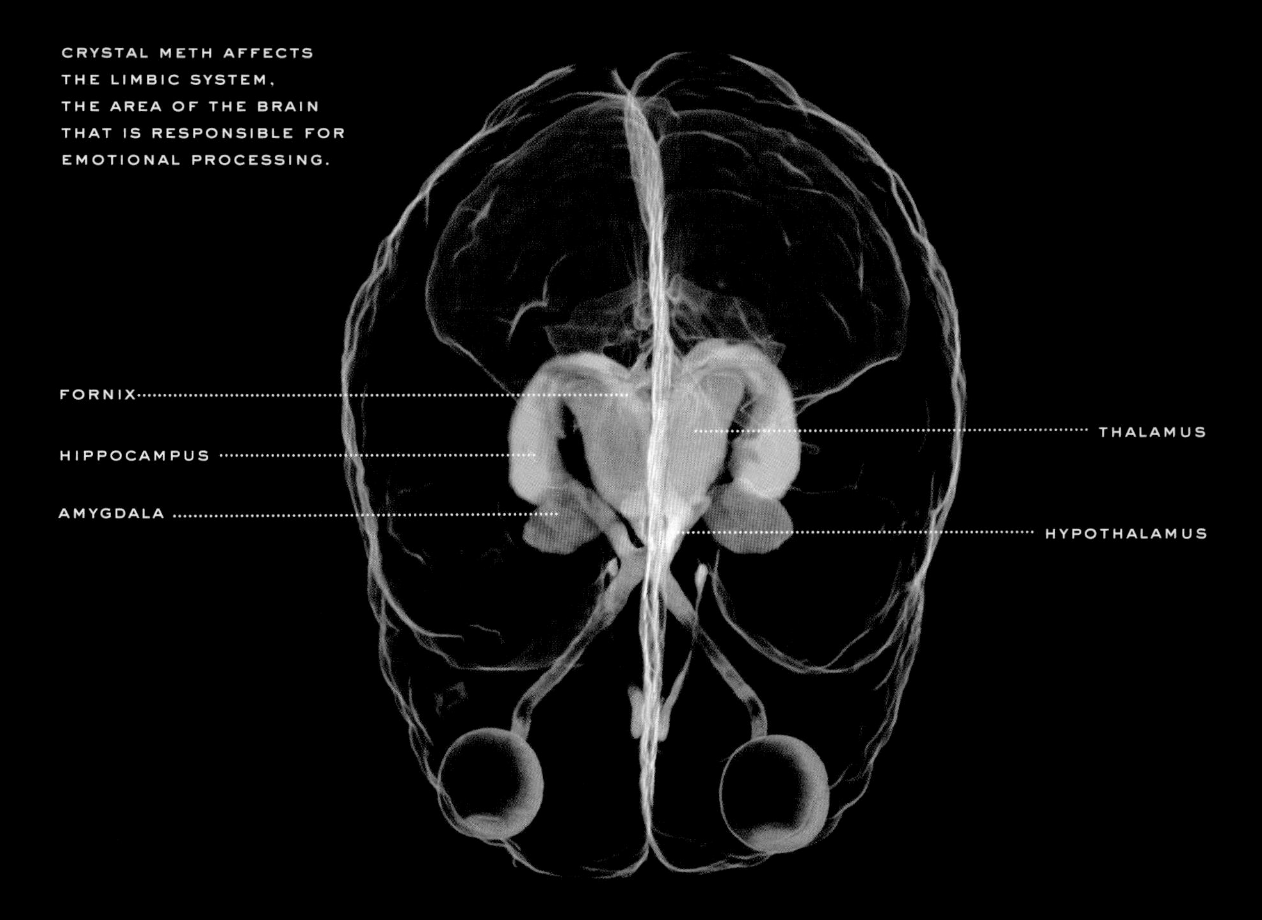

CRYSTAL METH AFFECTS THE LIMBIC SYSTEM, THE AREA OF THE BRAIN THAT IS RESPONSIBLE FOR EMOTIONAL PROCESSING.

Beyond the effects of attractiveness and two different measures of masculinity, the strongest predictor in a 2004 study of women's sexual permissiveness was the amount of money that they had spent on alcohol in the preceding month.[6]

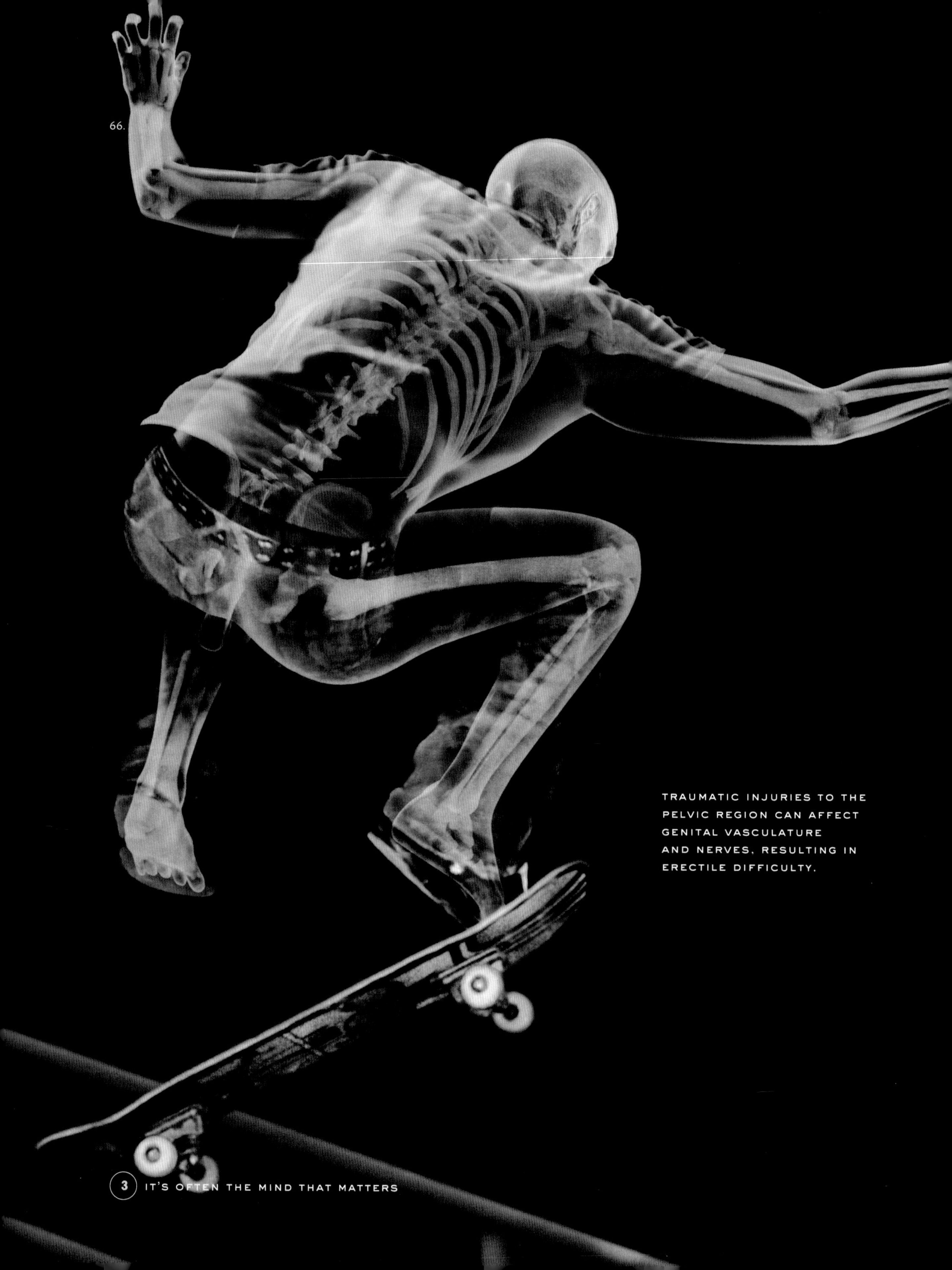

TRAUMATIC INJURIES TO THE PELVIC REGION CAN AFFECT GENITAL VASCULATURE AND NERVES, RESULTING IN ERECTILE DIFFICULTY.

REFLECTING ON REFLEXES

Although there is definitely a conscious component to arousal that is based in the CNS, there is also a component that is pure reflex. Some men and women with spinal cord damage can still experience physical arousal and orgasm even if they have no genital sensation. Sexual response can also be conditioned: the body can be retrained to respond to stimuli that it normally wouldn't perceive as sexual. These functions work through very different pathways, although their effects are both on the subconscious level. In the case of individuals with spinal cord injury, local nerve responses seem to be responsible for the ability to orgasm. The conditioned response, on the other hand, involves a change in brain activation where the mind itself is trained to respond in a sexual way to nonsexual stimuli.

Even though a serious sensation-destroying, or paralytic, spinal cord injury can leave a person still capable of sexual function, some much less severe injuries can demolish sexual capacity while leaving the rest of the body intact. Bicycle riding on an inappropriate saddle can, for example, lead to erectile dysfunction that can last from months to years. Other traumatic injuries to the pelvic region, like those that affect genital blood flow, vasculature, or nerves, can also cause erectile difficulty. So can penile fracture. In women, injury and scarring can lead to difficulty with penetration, as well as loss of sensation and interest in sex.

Peyronie's disease is a condition in which the connective tissue of the penis becomes scarred, possibly leading to penile curvature, shortening, disfigurement, or loss of capacity for erection. It's thought to occur in approximately three percent of older men, although numbers are probably much higher since it is an erectile dysfunction condition that is difficult for many men to discuss with their physicians. Although the exact cause of the disease is not known, it's thought to begin with some sort of trauma to the penis, possibly from damage occurring during vigorous sex. Then, if the damage doesn't heal successfully (perhaps due to an autoimmune response or excess inflammation), scar tissue can form and permanently injure the structure of the penis. Because of the nature of the disease, erectile dysfunction drugs and other medical treatments are not normally effective for resolving complications of Peyronie's disease. When erectile function is impaired or curvature becomes too extreme to allow for vaginal penetration, surgery is frequently the only option.

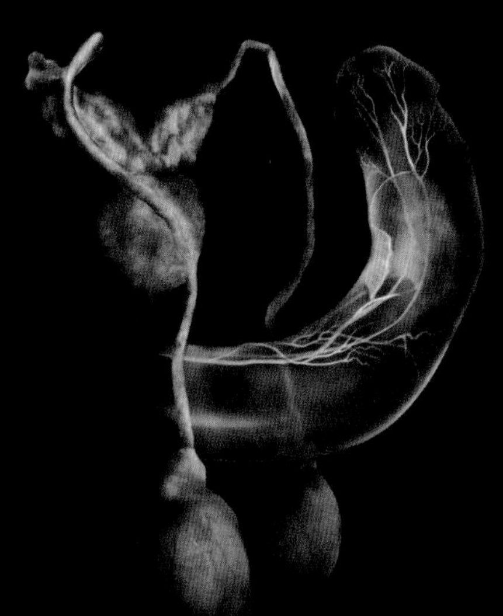

THE PENIS OF A MAN WITH PEYRONIE'S DISEASE MAY BECOME BENT DUE TO THE PRESENCE OF SCAR TISSUE.

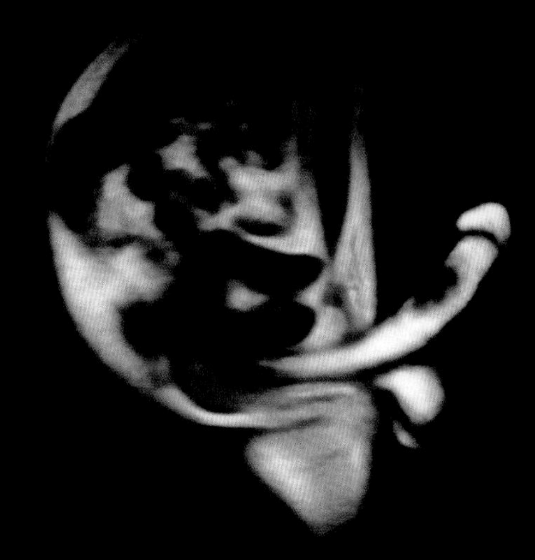

THIS MRI SHOWS THE POSITION OF THE PENIS DURING VAGINAL INTERCOURSE. IF THE PENILE TISSUE IS SCARRED AND BENT, IT MAY BE DIFFICULT TO ...ETRATION.

JUST AS EVERY PART OF YOUR BODY CAN BE INVOLVED IN THE SEXUAL RESPONSE, SO CAN EVERY PART OF YOUR BRAIN.

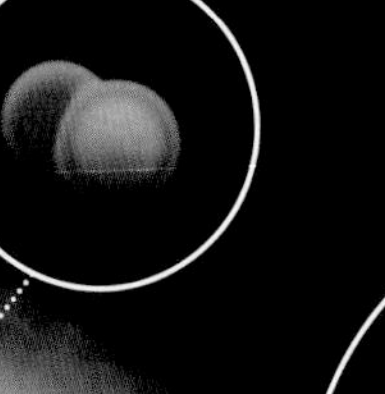

OXYTOCIN INCREASES TESTOSTERONE PRODUCTION IN BOTH MEN AND WOMEN; OXYTOCIN LEVELS SPIKE DURING ORGASM.

JUST THINKING ABOUT YOU. . .

Just as every part of your body can be involved in sexual response, so can every part of your brain. Sexual thoughts and fantasies arise in the outer part of the brain—the cerebral cortex—and hormones are controlled by the hypothalamus. But our emotional and instinctual responses originate elsewhere: in the limbic system.

The limbic system is a network of brain structures involved in the experience and expression of emotion. Depression and other emotional problems can be the result of damage to the limbic system, which is also known as the "pleasure center" of the brain. Sexual emotions, and our responses to drugs, gambling, and the other pleasures of life are all processed in the limbic system, and direct stimulation of these areas of the brain can lead to physical arousal.

Other parts of the brain are engaged in sexual processing as well. When, for example, men are exposed to erotic imagery, several parts of their brains become active. The thinking parts of their brain pay attention to the stimulus, while the limbic system and hypothalamus provide the motivation and activity needed for erection. Another part of the brain, the amygdala, seems to play an important role in coo dinating the sexual response, and some people believ it actually determines whether any given stimulus perceived as sexual. The amygdala is a component c the temporal lobe and,has been shown to play a role i the sexual responsiveness of many animal species. It intricately intertwined with the many different leve of the brain, from the cortex to the hypothalamus.

Although penile erection is basically a loca reflex, it is usually stimulated by sensations that ar processed in the brain. Smells, for example, can initiat sexual arousal, as can sights, sounds, or fantasies. On important area of the brain responsible for controllin erection is a specialized area of the hypothalamu known as the paraventricular nuclei (PVN). Nerve cel that start in the PVN make a hormone called oxytocir and send it directly into not only other areas of th brain, but also the spinal cord and bloodstream. Som of these nerves send oxytocin directly into the auto nomic pathways responsible for erection, taking muc less time to stimulate the penis then it would if th oxytocin had to circulate through the blood.

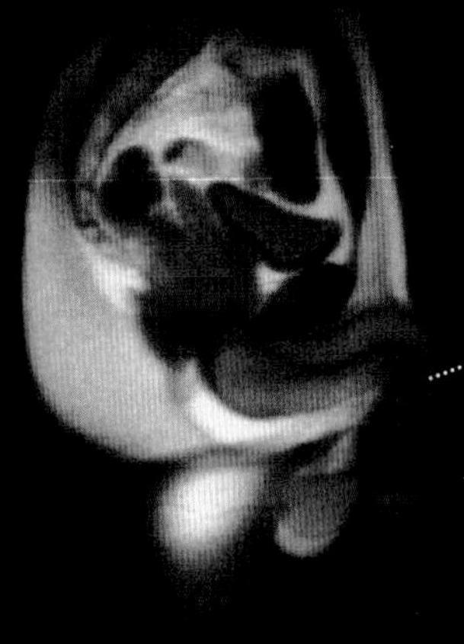

SEMI-ERECT TISSUE

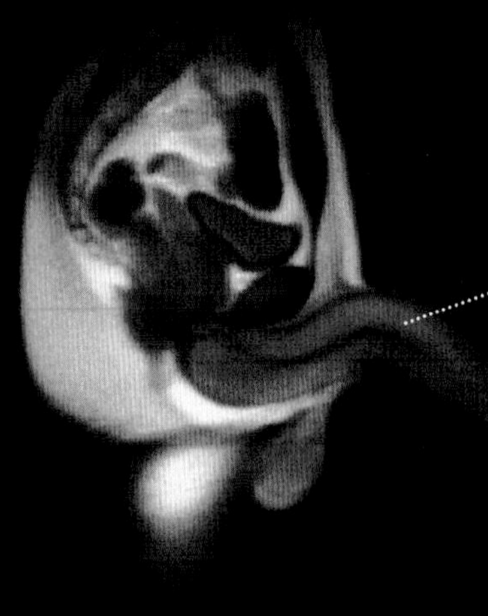

FULLY ERECT TISSUE

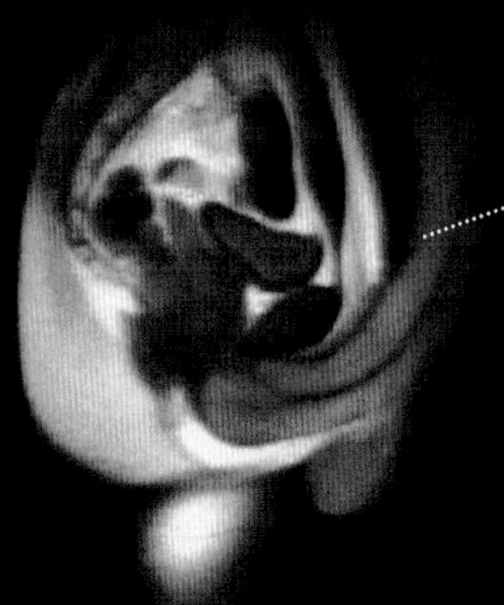

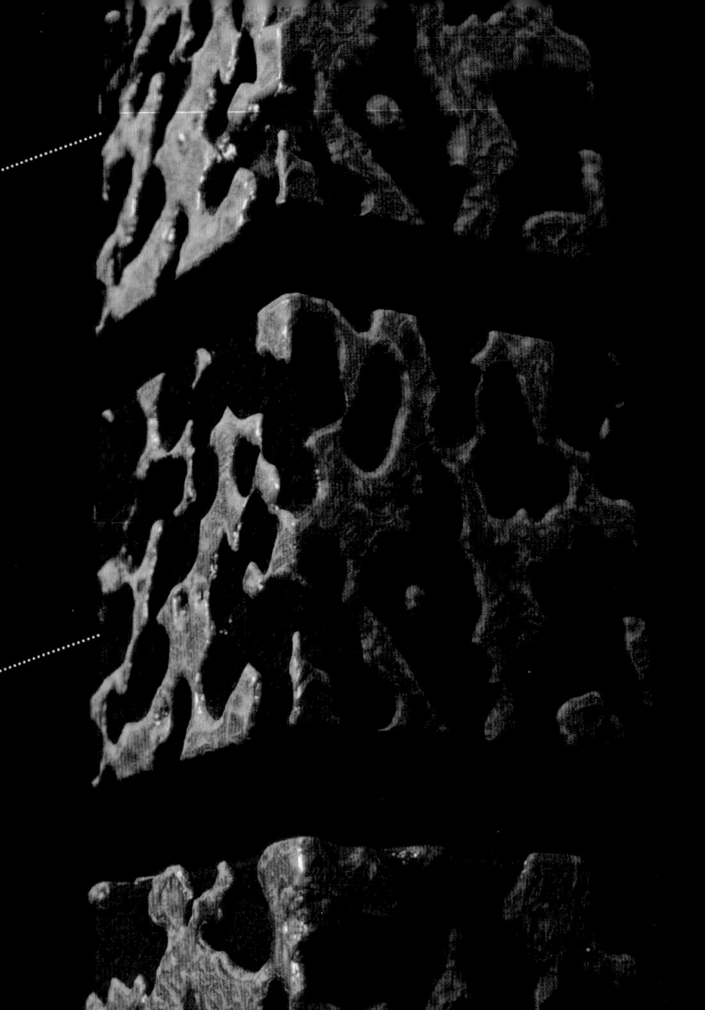

THE ERECTILE TISSUE OF THE PENIS CONTAINS MANY HOLLOW AREAS THAT CAN FILL WITH BLOOD.

If there is one chemical in charge of mediating sexual functioning, it is nitric oxide (NO). Although many chemicals are involved in the processes that control sexual arousal in men and women, nitric oxide plays an essential role in regulating blood flow. Blood flow is central to most of the obvious physical manifestations of the sexual response. Nitric oxide doesn't affect blood flow directly, but instead initiates a series of chemical reactions that allow the blood vessels to dilate. The most important milestones along this chemical pathway are two proteins known as cyclic GMP (cGMP) and PDE 5. cGMP is the protein that signals the muscles in the blood vessel walls to relax and fill with blood. It's critical that the amount of blood that flows into the affected vessels be strictly controlled, filling them to capacity—but not to bursting. That is the role of PDE 5. PDE 5 breaks down cGMP to prevent overstimulation of the blood vessels and excessive relaxation. When the system works, it provides an extremely effective mechanism to create and control arousal. However, problems start to occur if not enough cGMP is made or if cGMP is broken down too quickly. In either case, the blood vessels don't remain relaxed long enough to fill with sufficient blood to allow the penis to become erect, the clitoris to engorge, or vaginal lubrication to occur.

Current erectile dysfunction drugs work by blocking the effect of PDE 5 and allowing cGMP to remain active longer. These drugs can have beneficial sexual effects for men—and, some studies have found, for women. They make it easier for men to both achieve and maintain an erection, and for women to experience the vaginal lubrication and the enhanced clitoral sensitivity associated with normal sexual arousal.

JUST SAY NO

SMOOTH PENILE MUSCLE RELAXED

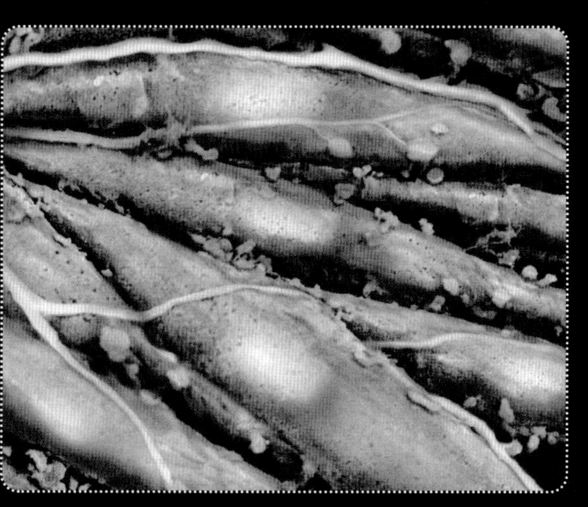

SMOOTH PENILE MUSCLE CONTRACTED

ERECTION IS CAUSED WHEN NITRIC OXIDE RELEASED FROM LOCAL NERVE ENDINGS STARTS A CHEMICAL CASCADE THAT CULMINATES IN THE RELAXATION OF THE SMOOTH MUSCLE OF THE BLOOD VESSEL WALLS AND A VAST INCREASE IN LOCAL BLOOD FLOW.

A LITTLE TOO SWEET

Diabetes can have serious sexual side effects in men, including decreased libido, difficulty achieving orgasm, and retrograde ejaculation. The most well-known sexual complication of diabetes, however, is erectile dysfunction, which occurs in between 20 percent to 70 percent of diabetic men. Studies that have compared diabetic to non-diabetic men find that men with diabetes are almost twice as likely to have problems with erection.

Diabetic men are at increased risk for erectile dysfunction because of, among other reasons, the damage to nerves that diabetes causes throughout the body. When this damage occurs to the parasympathetic nerves of the penis, it directly affects local sensation and function. The less well controlled a patient's blood sugar levels, and the longer they've had diabetes, the more likely they are to experience nerve damage and erectile dysfunction.

Diabetic men can also have problems with their NO cascade. Not only may it be difficult for their brains to stimulate release of NO because of nerve damage, but diabetics also have increased production of arginase, an enzyme found in the liver. This can make it difficult for them to make enough NO, because arginine is a necessary ingredient in NO production and too much arginase in the body can use it up.

The blood vessel damage that is so well known in diabetics can also cause erectile dysfunction. In uncontrolled diabetes, poorly managed blood flow can cause the smooth muscle of the penis to be replaced by less effective fibrous muscle. If enough fibrous muscle is present, it can make it impossible for a man to have an erection without a penile implant. Blood vessel changes in diabetic men can also reduce the volume of blood that enters the penis, and without sufficient blood flow it may be impossible to get or maintain an erection. One study found that more than 95 percent of diabetic men have impaired blood flow in their penile arteries.

Unfortunately, because of the types of physical damage associated with diabetes, standard medical treatments for erectile dysfunction are frequently less effective in diabetics than they are in other men. Researchers continue to investigate new treatments for this high-risk population.

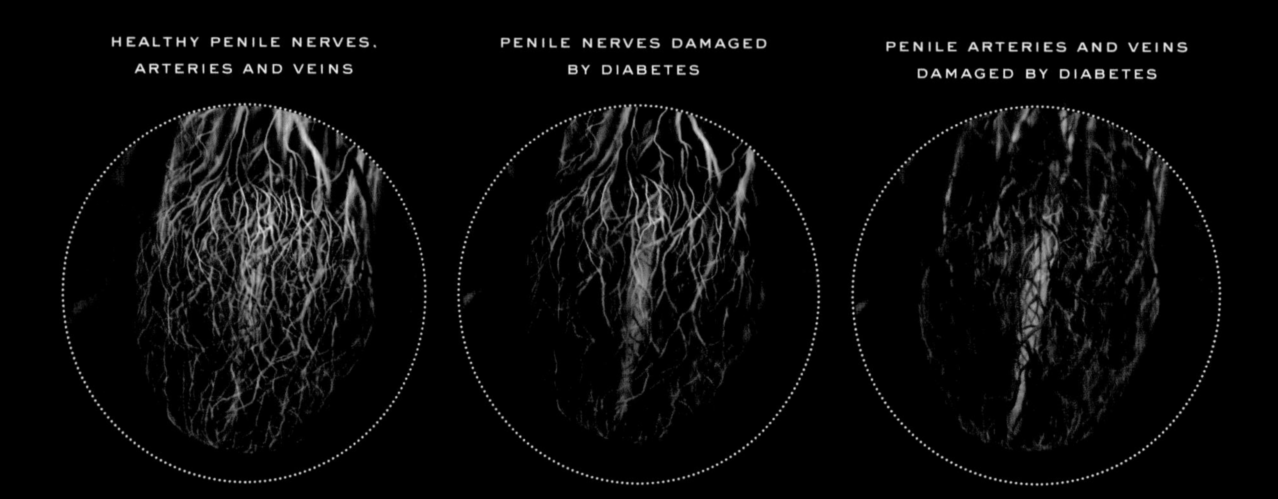

Women with diabetes may also experience sexual health difficulties, including reduced desire, decreased capacity for arousal, and painful intercourse. Decreased vaginal lubrication, however, is the most significant component of sexual dysfunction in diabetic women. It is probably caused by vascular damage similar to that seen in men. Without sufficient blood flow into the vagina, there is insufficient transudation of fluid, and vaginal lubrication is reduced.

There has been relatively little quantitative research on sexual dysfunction in women with diabetes, although there has been intense study of diabetic men. The research that has been done suggests that between 20 percent to 70 percent of women with type I diabetes and approximately 40 percent of women with type II diabetes have some degree of sexual dysfunction. Figures vary widely, but studies have generally found that the prevalence of dysfunction in diabetic women is about twice that of non-diabetic women from the same population.

Some scientists have suggested that this data may be complicated by body-image issues caused by cofactors of diabetes, like age and obesity. In other words, because type II diabetes is found in older, heavier women, it's hypothesized that biology may be less of the problem than loss of attractiveness. Can you imagine anyone suggesting that being bald or fat is an important cause of sexual dysfunction in men? This is just additional proof that further research is needed to clarify the nature of sexual dysfunction in diabetic women, so that appropriate treatment regimens can be developed. One thing, however, is clear: the more complications of diabetes a woman has, the more likely she is to have difficulties with her sex life.

SUGAR AND SPICE MAY NOT BE SO NICE

THE ERECTILE TISSUE OF THE CLITORIS AND THE MAJOR MUSCLES OF THE PELVIS

HEALTHY CLITORAL AND LABIAL NERVES, ARTERIES, AND VEINS

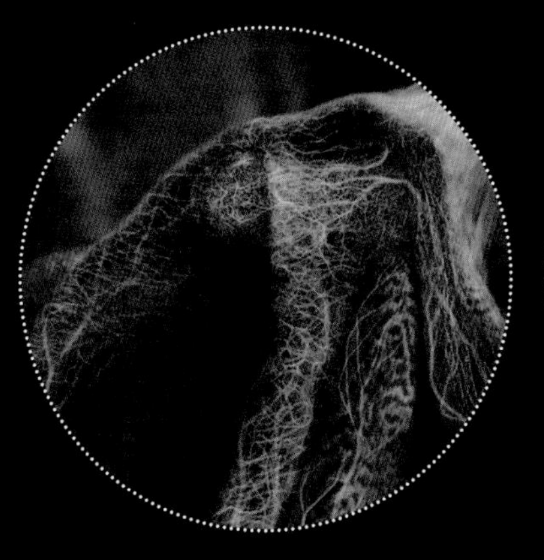

CLITORAL AND LABIAL NERVES DAMAGED BY DIABETES

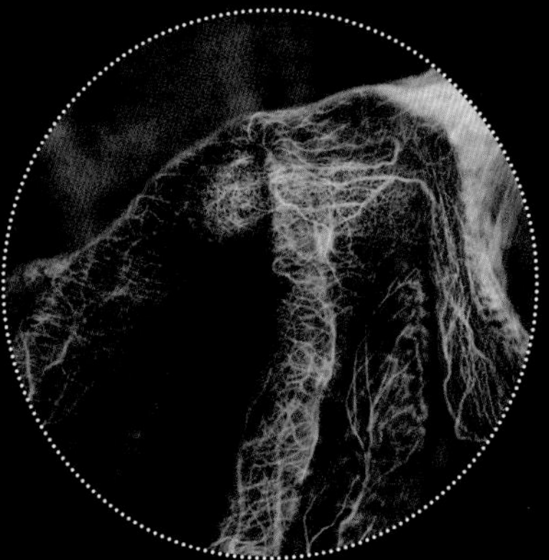

CLITORAL AND LABIAL ARTERIES AND VEINS DAMAGED BY DIABETES

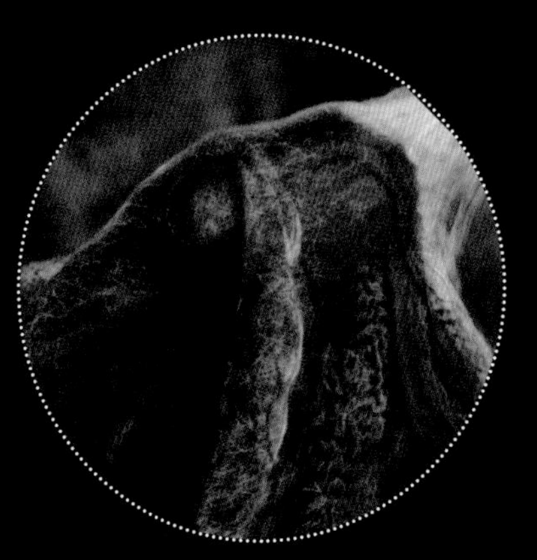

OBESITY AND SUBSTANCE ABUSE ARE ASSOCIATED WITH SEXUAL DYSFUNCTION.

UNDERPRODUCTION OF THYROID HORMONE MAY INCREASE THE RISK OF OBESITY.

PLAQUE HAS BEGUN TO BUILD UP IN THIS BLOOD VESSEL, REDUCING BLOOD FLOW.

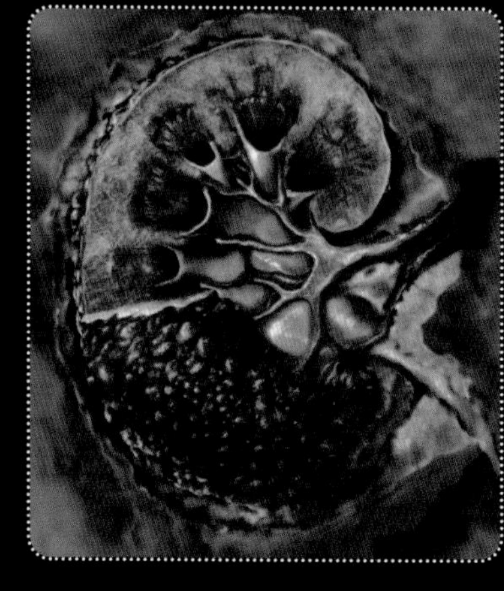

KIDNEY DISEASE ALTERS GONADAL FUNCTION AND HORMONE

NOT FEELING UP TO ANYTHING?

Obesity and the health problems associated with it—vascular disease, hypertension, high cholesterol, and diabetes—are all linked to sexual dysfunction in both men and women. These diseases are associated with a process called endothelial dysfunction, a disruption of the interior layer of the blood vessels, which can affect the nitric oxide cycle in the body. In fact, because of the association between endothelial dysfunction and serious cardiovascular illness, some scientists have suggested that erectile dysfunction may be a useful warning sign for the early stages of heart disease in men.

Besides their effects on the nitric oxide cycle, these conditions and their treatments all have negative effects on systemic circulation. If blood vessels are too hardened to dilate, physiological arousal is impossible. Anti-hypertensive drugs can control high blood pressure, but they may make it difficult to attain the local increases in blood pressure necessary for erection and lubrication. Anything that strains the cardiovascular system also puts strain on the mechanics of sexual function.

Chronic diseases can also have debilitating effects on sexual function. They can affect the brain, the vasculature, and the nervous system, and can have direct effects on sexual function. Nerve damage from multiple sclerosis can affect sexual function by causing two opposite effects. It can create a reduction in genital sensation and a loss of interest in sexual activity; conversely, it can cause a hypersensitivity to genital stimulation, making sex uncomfortable to the point of pain. Kidney disease, in contrast, affects a person's sex life by altering both gonadal function and hormone production in the brain. When the kidneys are no longer able to filter the blood properly, it becomes contaminated with components that should be excreted in the urine. This can cause mixed signals affecting the hypothalamic and pituitary control of hormone synthesis, and it can also directly impair function of the ovaries and testicles. Furthermore, since high blood pressure is a common correlate of kidney disease, many of the sexual issues found in individuals with hypertension are also found in those with kidney disease.

Cancer and its treatments also have multiple deleterious effects on sexual function. The pain associated with the disease, and the massive discomfort, nausea, and fatigue common to its treatment, can reduce sexual interest and interfere with arousal. The physical changes that cancer patients experience, from hair loss to amputation, can also drastically affect self-image and diminish their interest in sex. Cancer patients may be unwilling to let a partner witness the effects of the disease on their body. Reproductive cancers can be particularly devastating to sexual function, due to changes in sensation or function and even loss of reproductive organs.

All in all, any illness that affects a person's ability to function in daily life is also likely to have negative effects on his or her sexual life. Expanding the definition of sex beyond the standard repertoire of genital sensation may allow people with chronic diseases to find greater happiness in their sexual activities and improve their overall quality of life as well.

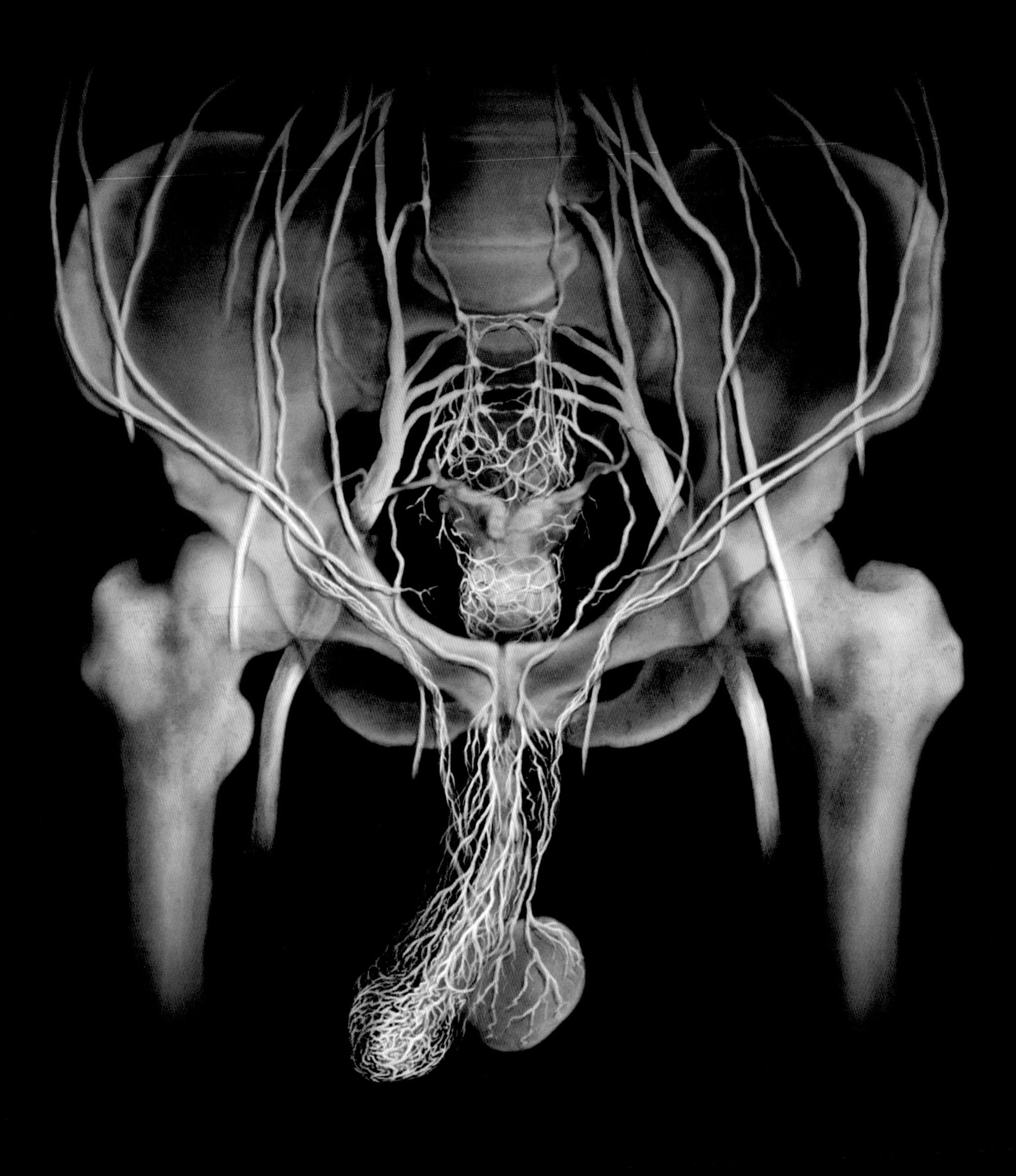

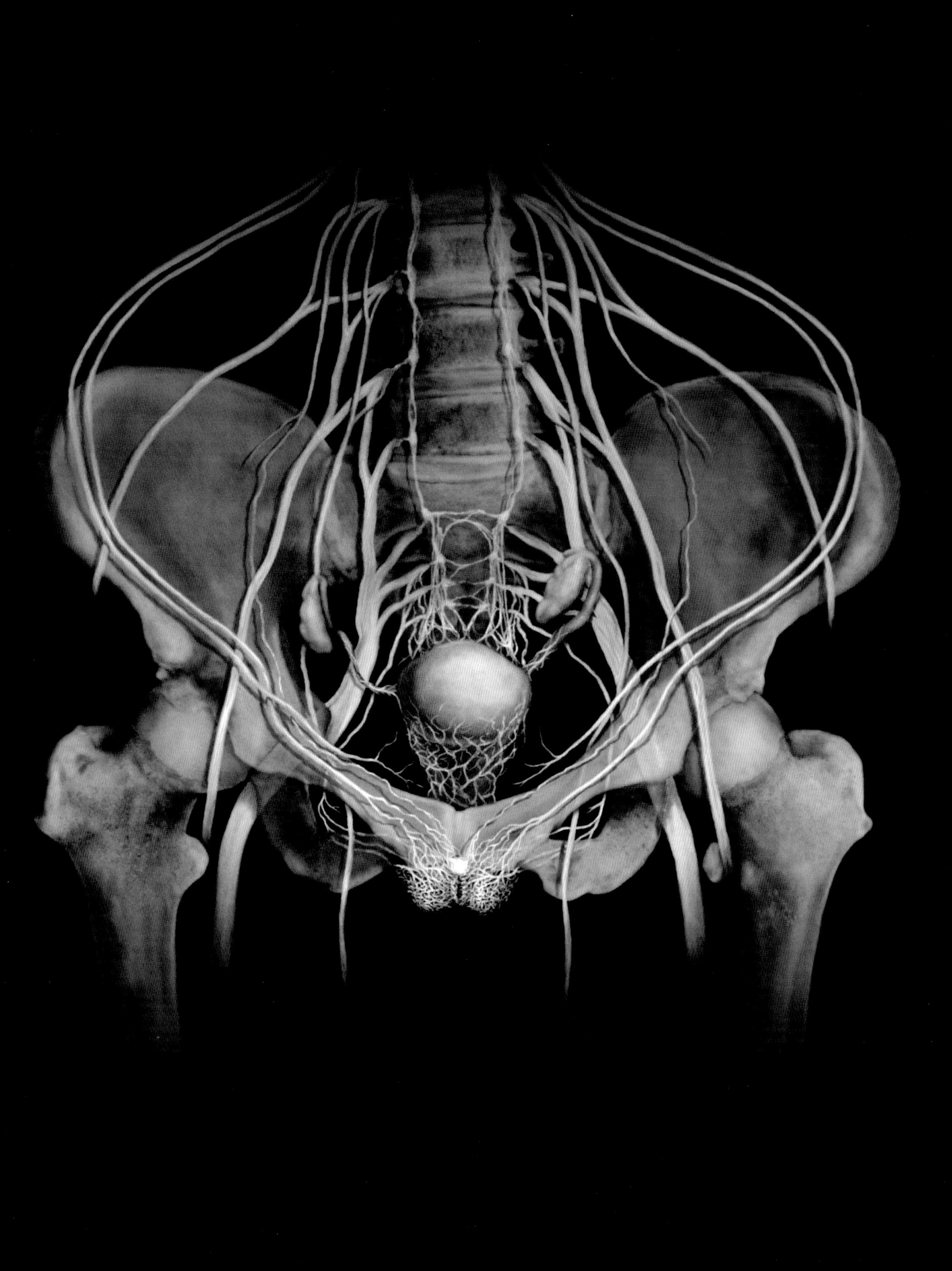

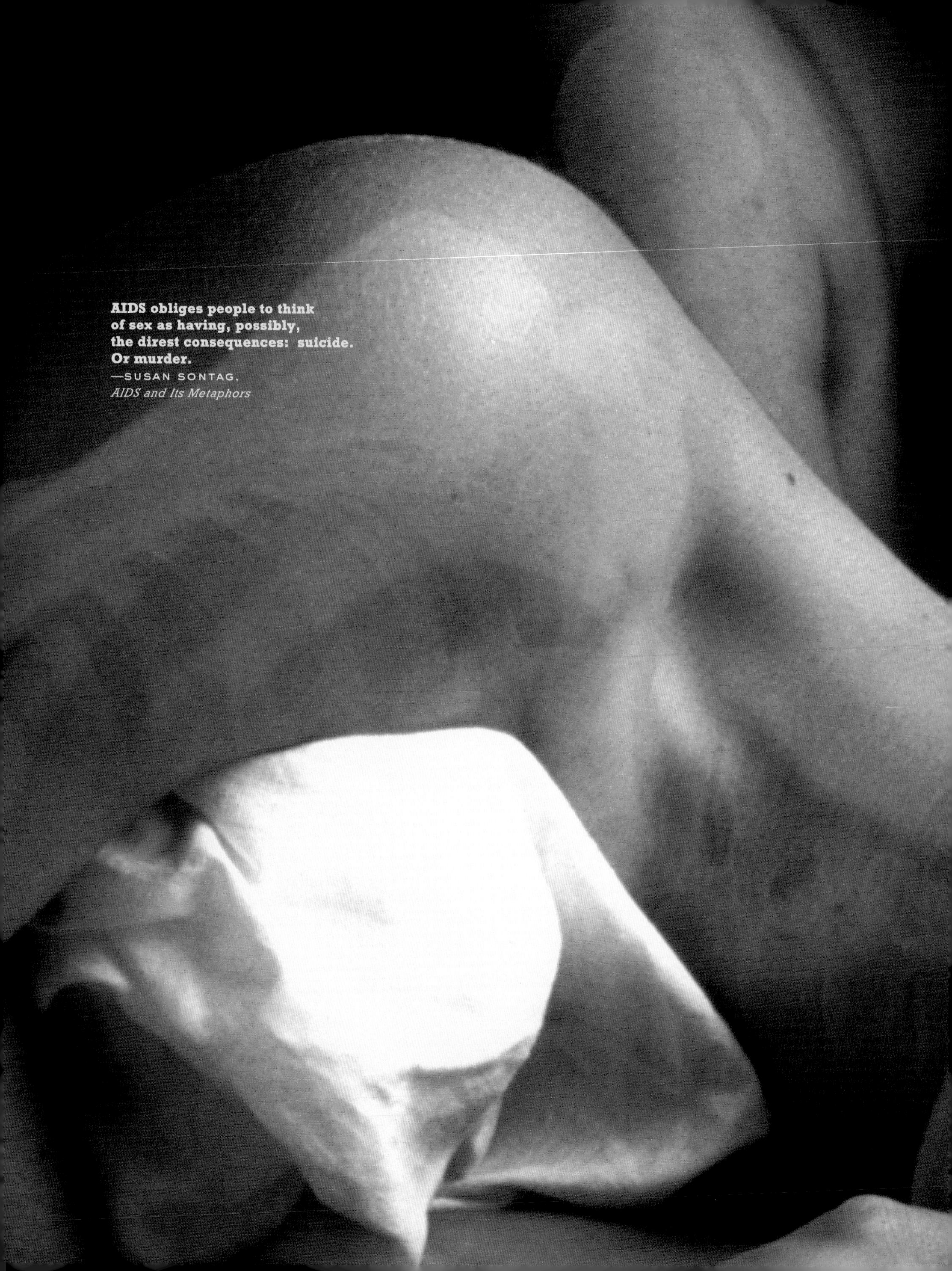

AIDS obliges people to think of sex as having, possibly, the direst consequences: suicide. Or murder.

—SUSAN SONTAG, *AIDS and Its Metaphors*

WHEN THINGS GO WRONG: DISEASES AND DISORDERS

Sex isn't always a healthy choice. Impulsive decisions can lead to lifelong problems: serious disease and infertility. The fact that more than three quarters of the sexually active population has at least one sexually transmitted disease (STD) is a sign of the failure of society to educate people to make smart choices. It's especially disheartening since most STDs are preventable through simple behavioral changes—changes that can save your life, preserve your ability to have children, and make the sexual experience more enjoyable, because you won't be worrying about the possible consequences.

LET'S HAVE A CONVERSATION

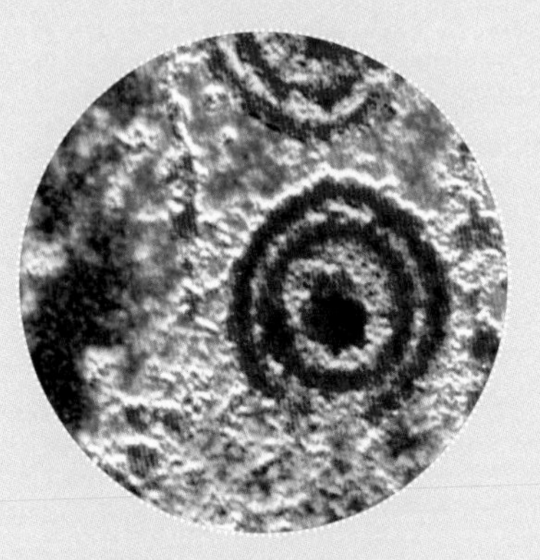

YOUR CHOICE—ALWAYS: Your sexual choices should not be spur-of-the-moment decisions, because making the wrong ones can have devastating effects on the rest of your life. You never have to have sex with someone else, no matter how desperate you feel or how much pressure is put on you. In case of emergency, smash open the box of sexual aids and enjoy *yourself* instead.

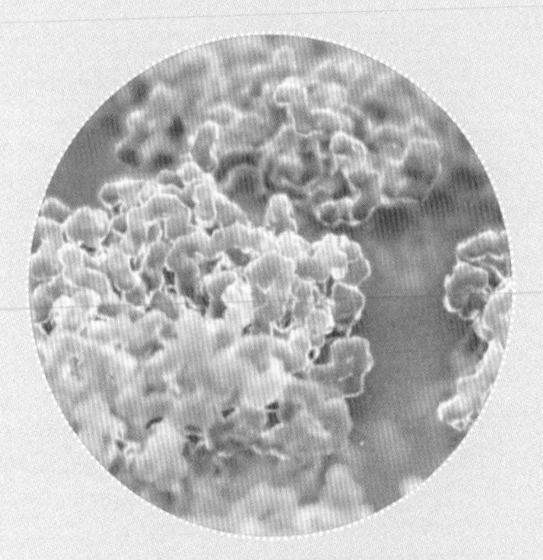

I CAN CATCH CANCER?: The "Big C" stands as much for contagious as it does cancer. Some forms of reproductive cancer are actually sexually transmitted diseases.

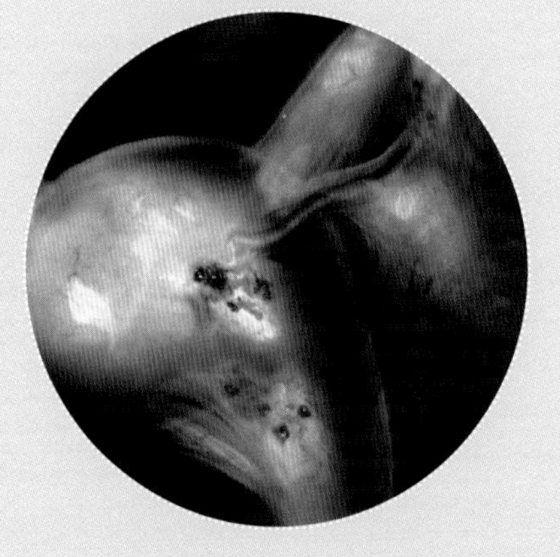

IT'S WHAT'S INSIDE THAT MATTERS: Breast, endometrial, and ovarian cancers are terrifying diseases that strike at the heart of womanhood, but they can be treated if caught early enough. Loving, and paying attention to, your body not only makes you feel beautiful, but helps you to stay alive.

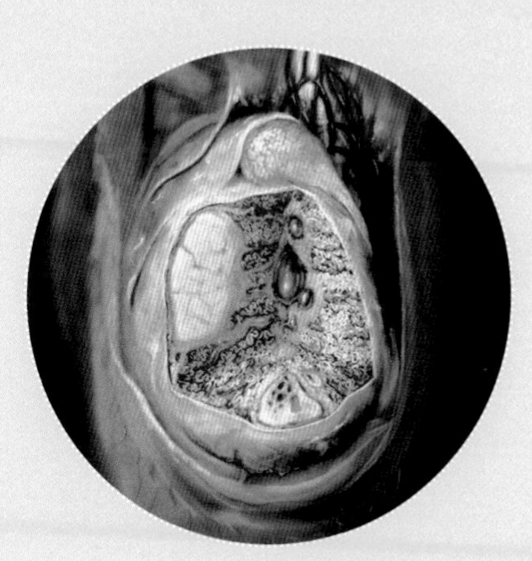

TOUCH MYSELF WHERE?: Touching yourself isn't just fun, it's important for your health. Self-examination during masturbation can help men catch the early signs of testicular cancer while it is still treatable. Male reproductive cancer risk starts early and extends all the way into old age. For women, regular breast self-examination can help detect breast cancer in its early stages as well, greatly increasing chances of survival.

DON'T SCRATCH THIS ITCH: Genital pain and discomfort occur in both men and women for many different reasons. Some of them are serious, some of them are minor, but all of them should be evaluated by your physician. In the meantime, listen to your mother and don't scratch. It really does only make things worse.

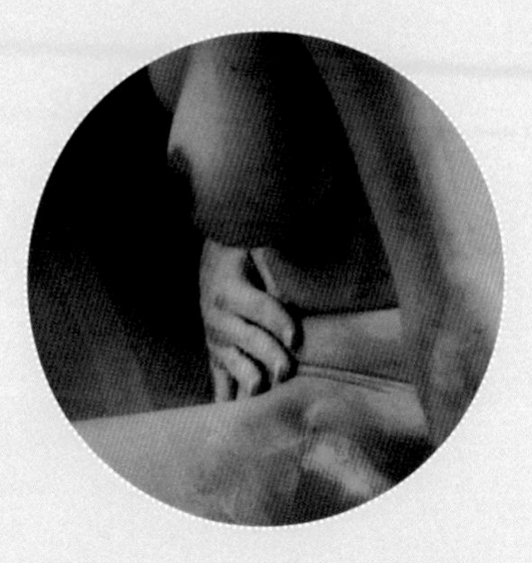

Most of the bacterial diseases that can be transmitted sexually are curable. . .if they're found in time. Unfortunately, a majority of STDs are asymptomatic—that is, there are no obvious external symptoms—and people often have no idea they have them until years after they've been infected. Those years of ignorance not only provide an opportunity for them to spread the disease to others, but also for the disease to wreak havoc on their reproductive tracts and sometimes elsewhere in the body. On a positive note, all of the side effects of bacterial STDs can be prevented by consistent condom use for both penetrative and oral sex, since some of these diseases can also be transmitted through the mouth. (Note: Treatments for many of the diseases covered in this chapter are discussed in Chapter Five.)

Different sexually transmitted diseases lurk in different parts of the reproductive tract, depending on where they "like" to do their damage. Protozoa loiter in the vagina, infecting anyone they can get their

DON'T PICK UP HITCHHIKERS

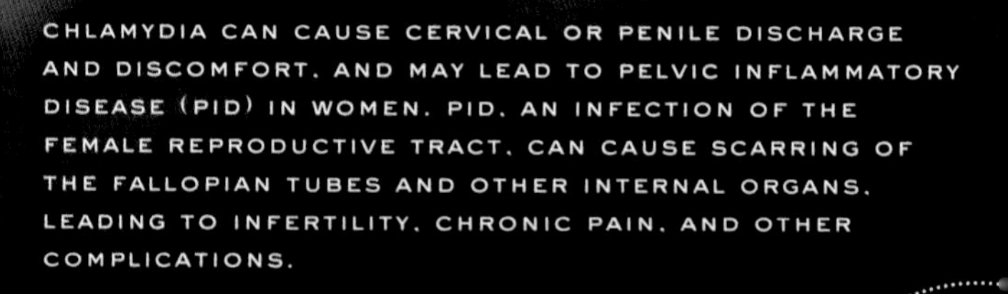

CHLAMYDIA CAN CAUSE CERVICAL OR PENILE DISCHARGE AND DISCOMFORT, AND MAY LEAD TO PELVIC INFLAMMATORY DISEASE (PID) IN WOMEN. PID, AN INFECTION OF THE FEMALE REPRODUCTIVE TRACT, CAN CAUSE SCARRING OF THE FALLOPIAN TUBES AND OTHER INTERNAL ORGANS, LEADING TO INFERTILITY, CHRONIC PAIN, AND OTHER COMPLICATIONS.

cilia on. Higher up around the cervix is where the bacteria—chlamydia and gonorrhea—go to steal young women's fertility. Finally, there's syphilis, one of the deadliest of the bunch, waiting for its chance to pounce.

Chlamydia and gonorrhea are two of the most commonly diagnosed sexually transmitted diseases. Both are highly asymptomatic—more than 96 percent of individuals with chlamydia and approximately half of individuals with gonorrhea have no symptoms. Even when they cause no symptoms, chlamydia and gonorrhea increase the risk of acquiring other infections such as HIV. In addition, if left untreated, these infections can lead to infertility and other serious health consequences. Most cases of these diseases occur in people in their teens and twenties, in part because women of younger ages are more susceptible due to differences in their cervix; a young woman has a more delicate cervix than an older woman, making her easier to infect.

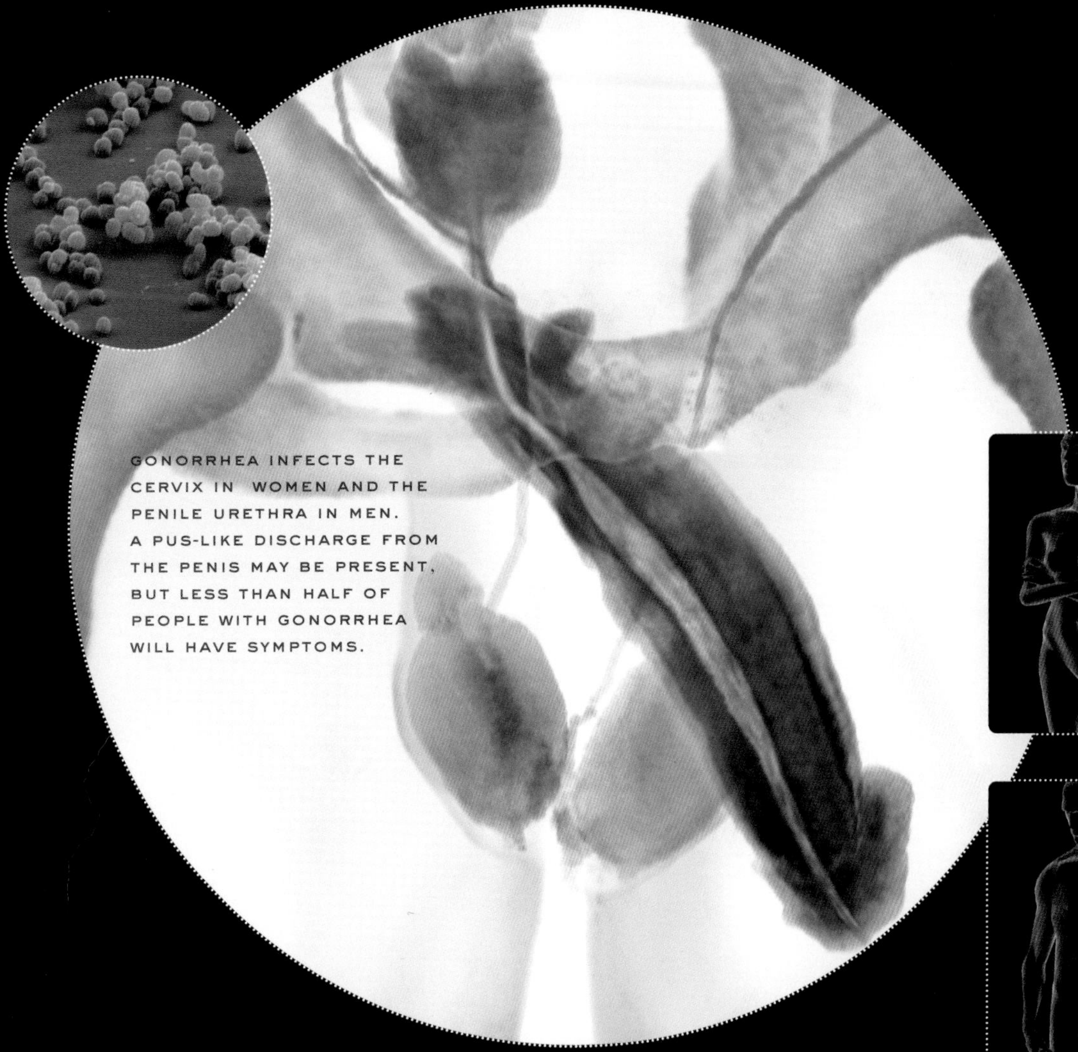

GONORRHEA INFECTS THE CERVIX IN WOMEN AND THE PENILE URETHRA IN MEN. A PUS-LIKE DISCHARGE FROM THE PENIS MAY BE PRESENT, BUT LESS THAN HALF OF PEOPLE WITH GONORRHEA WILL HAVE SYMPTOMS.

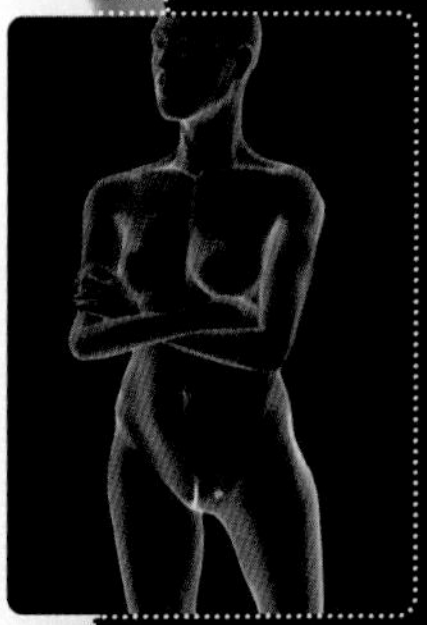

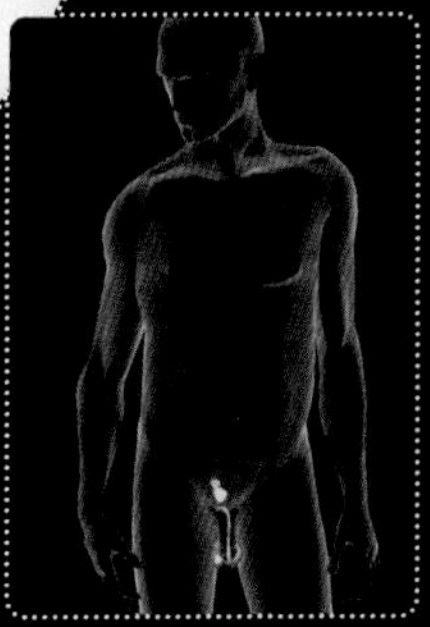

WHAT DID CHRISTOPHER COLUMBUS, ABRAHAM AND MARY TODD LINCOLN, AND VINCENT VAN GOGH HAVE IN COMMON?

They all probably had syphilis.

Syphilis is a very nasty disease that can do very serious damage to the human body—including the brain—if left untreated. Unlike the other curable STDs, it causes not just a localized infection, but a systemic one. In its first stage, it produces genital ulcers. In its later stages, after the ulcers have healed, it can damage most of the body's systems. Numerous famous people have suffered from syphilis, and some historians have wondered if it played any role in their sometimes erratic behavior. It can, after all, have some pretty strange personality effects.

Trichomonas vaginalis is one of the most common sexually transmitted diseases in the United States. It's caused by a protozoa—a small one-celled organism with a four-stranded tail—and it is relatively easy to identify during a medical exam. Interestingly, it's the only one of the curable STDs for which infection becomes more common with increasing age. Just like chlamydia and gonorrhea, it's associated with increased risk for HIV, as well as multiple health complications, and it can even affect pregnancy.

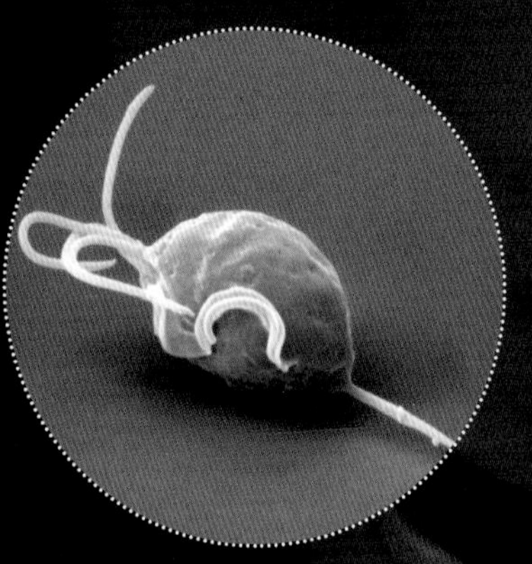

TRICHOMONAS IS USUALLY THOUGHT OF AS A VAGINAL INFECTION. ALTHOUGH IT CAN INFECT THE PENILE URETHRA, IT RARELY CAUSES PROBLEMS IN MEN.

Perhaps more than any other disease before or since, syphilis in early modern Europe provoked the kind of widespread moral panic that AIDS revived when it struck America in the 1980s.

—PETER LEWIS ALLEN,
The Wages of Sin: Sex and Disease, Past and Present

SYPHILIS IS AN STD CAUSED BY A BACTERIUM CALLED *TREPONEMA PALLIDUM*.

IN THE PRIMARY STAGE SYPHILIS CAUSES A FIRM ULCERATED NODULE. IN ITS LATER STAGES IT BECOMES SYSTEMIC AND CAUSES DAMAGE THROUGHOUT THE BODY.

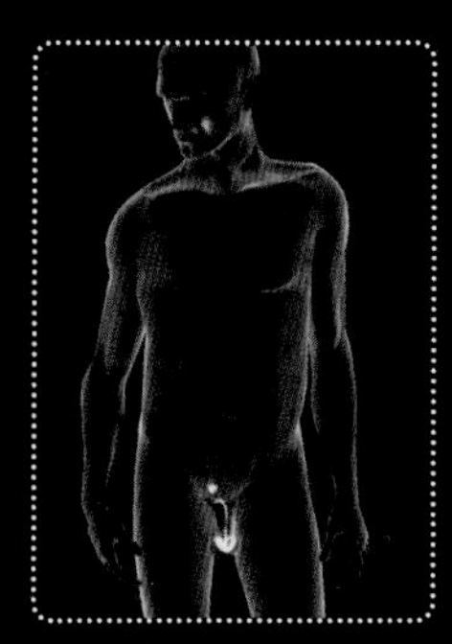

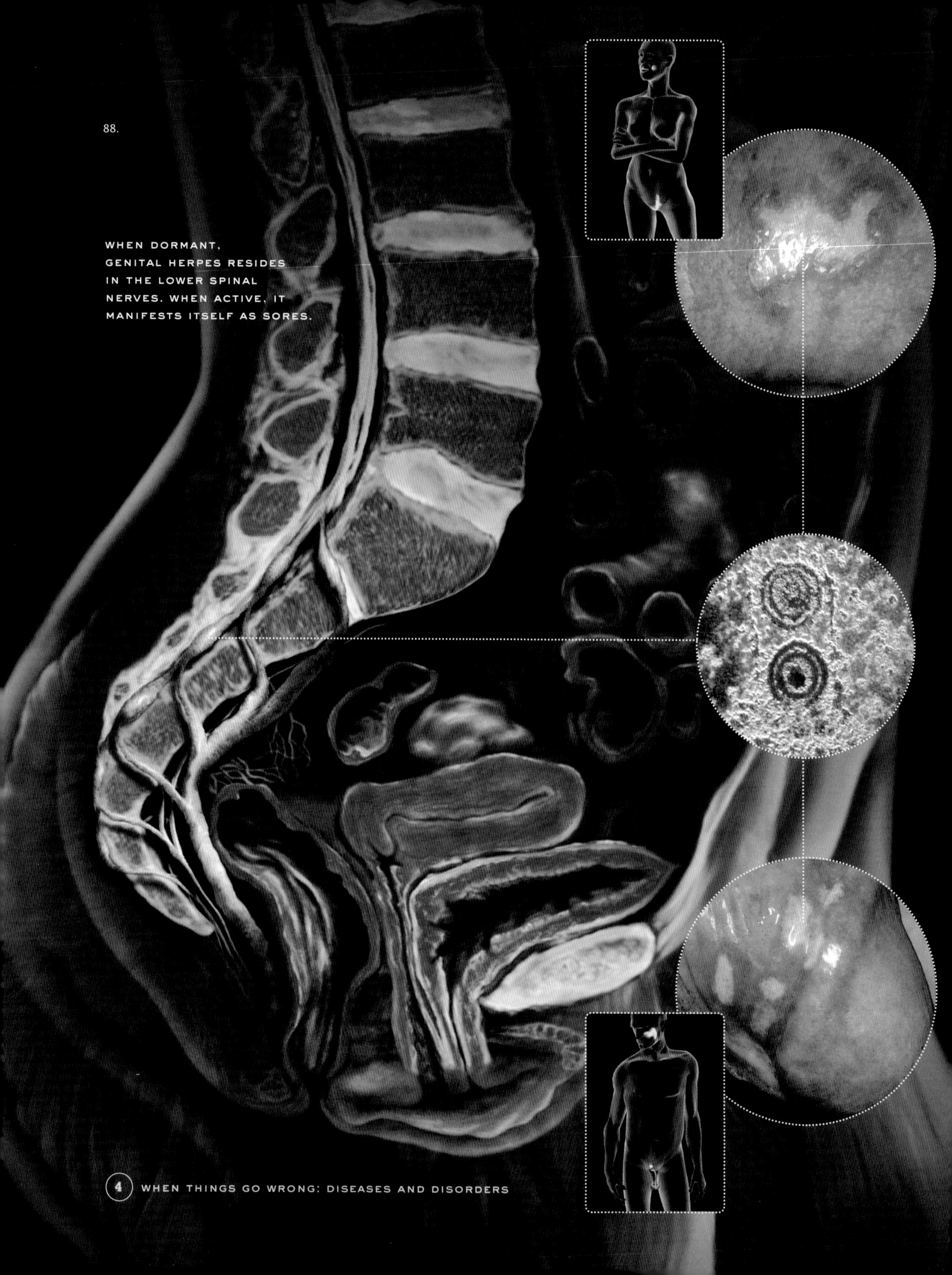

WHEN DORMANT, GENITAL HERPES RESIDES IN THE LOWER SPINAL NERVES. WHEN ACTIVE, IT MANIFESTS ITSELF AS SORES.

HSV, HPV, HIV—HELP!

The three most common currently incurable STDs are all caused by viruses. They don't, however, have very much in common, other than very similar three-letter acronyms.

A person who has herpes (HSV), may actually have one of two different conditions. Oral herpes is usually caused by HSV 1. This is the virus that causes cold sores, although many people have no symptoms. 50 percent to 80 percent of the U.S. population is positive for oral herpes—they normally acquire it as children from kissing their relatives. Genital herpes is usually caused by HSV 2, and it causes similar sores in the genital region. Almost one in four Americans is positive for genital herpes, although it's asymptomatic in 90 percent of cases. It's important for people to know, however, that herpes can still be contagious even when no sores are present and that oral sex can transmit herpes from the mouth to the genitals and vice versa.

Human papilloma virus (HPV) is the virus responsible for both genital warts and several reproductive cancers. Virtually all, if not all, cases of cervical cancer are caused by HPV, as are many cases of anal, rectal, and penile cancer. There are more than 70 types of HPV, but fewer than 30 are transmitted sexually and only a small subset of those are cancer-causing. Unfortunately, even consistent condom use cannot completely prevent the spread of HPV, since it is spread by skin-to-skin contact, and not solely by secretions. This may be why more than 75 percent of the population has been infected with at least one type of HPV. There is hope, however, since vaccines to prevent the most dangerous strains of the virus will soon be available.

HUMAN PAPILLOMA VIRUS (HPV) INFECTS THE GENITAL SKIN.

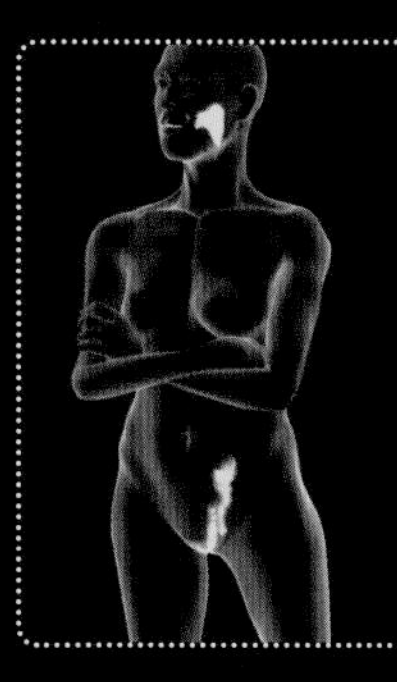

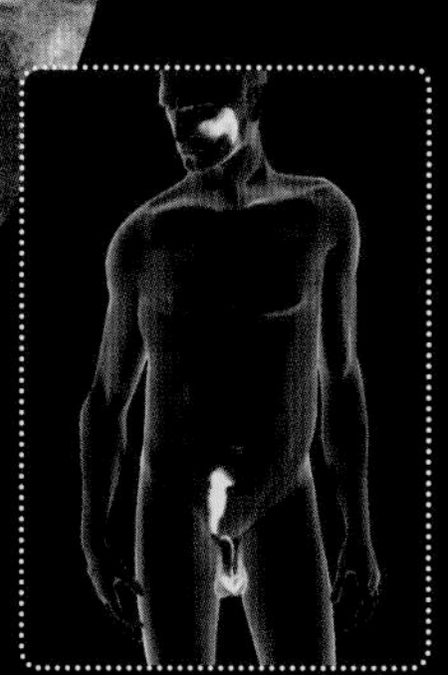

The great unspoken on the heterosexual AIDS front has been how behavior is still determined by the old psychosexual minuet of the sexes, the lack of responsibility in young men and of assertiveness in young women.
—ANNA QUINDLEN,
The New York Times

HIV IS A RETROVIRUS THAT IS TRANSMITTED THROUGH BLOOD, SEMEN, VAGINAL SECRETIONS, AND OTHER BODILY FLUIDS. THE VIRUS USES THE BODY'S CELLULAR MACHINERY TO REPLICATE ITSELF AND SPREAD.

HIV: VERY DANGEROUS, VERY PREVENTABLE

Human immunodeficiency virus (HIV) is the virus that causes AIDS. It is not a disease limited to gay men, injection drug users, or the morally weak. It is not a plague sent down by God. It is a terrible virus that is devastating the people of the world. It is also highly preventable. If everyone followed safer sex and universal precautions for infectious disease, there would be no new infections caused by this fragile, difficult-to-transmit virus. Unfortunately, being safe is not always easy. Some individuals are powerless to negotiate condom use, and others recognize it would be dangerous to try. Others don't want to risk their sexual enjoyment by using latex barriers for oral, manual, anal, or vaginal sex. Heterosexual women are the fastest-growing group of infected individuals in much of the world, because many cannot, or will not, ask for safer sex from their male partners. In places and situations where women are valued only for their child-bearing capabilities, a woman who insists on using a condom is viewed as being useless. More than that, she's viewed as not trusting her mate, which can cost her her relationship with the man who may be her sole source of support. It may even cost her her life.

Although infection with HIV is no longer a death sentence, it is still an incredibly serious disease. HIV is dangerous because it weakens the immune system, which makes infected individuals susceptible to many types of diseases that healthy people could easily avoid. Those infections are what kill a person with AIDS, not the virus that has made them so easy to infect. HIV is very difficult to treat. Although the drugs are constantly improving, the treatment regimens are difficult to follow, and many of the drugs have side effects that can feel far worse than the disease. And in many developing countries, the drugs may be too expensive for people to afford.

HIV OR AIDS?

To be diagnosed with AIDS, a person must be HIV-positive and must either have a CD4 (type of immune cell) count below 200 cells/mm^3, or one of the AIDS-defining conditions below. It is possible to go from having AIDS back to simply being HIV-positive if a person's immune system rebounds or if he or she recovers from the defining infection.

AIDS-defining conditions:

- Candidiasis—yeast infection
- Cervical cancer (invasive)
- Coccidioidomycosis—a type of fungal infection
- Cryptococcosis—a type of fungal infection
- Cryptosporidiosis—a severe diarrheal disease caused by a water-borne parasite
- Cytomegalovirus disease—a type of viral infection
- HIV-related encephalopathy—alteration of the brain
- Herpes simplex (severe infection)
- Histoplasmosis—a type of fungal infection
- Isosporiasis–a severe diarrheal disease caused by a protozoa
- Karposi's sarcoma—cancer with characteristic purple lesions
- Lymphoma (certain types)—cancer of the lymphatic system
- Mycobacterium avium complex—a type of bacterial infection
- Pneumocystis carinii/jiroveci pneumonia–specific types of bacterial pneumonia
- Recurrent pneumonia
- Progressive multifocal leukoencephalopathy—destruction of the covering of the nerve cells
- Salmonella septicemia (recurrent)—a severe form of bacterial diarrhea
- Toxoplasmosis of the brain—a type of parasitic infection
- Tuberculosis—an infectious disease caused by the tubercle bacillus
- Wasting syndrome—a disease characterized by unintentional weight loss and weakness

Cancer is uncontrolled and dangerous cell growth. When the mechanisms that normally keep cell division under control fail, cells can divide too often, damage and invade surrounding tissue, and cause sickness and death. The "C-word" will touch almost every family, and it is terrifying to many people at a deep, instinctual level.

Cervical cancer is highly preventable. If she gets regular Pap smears, a diagnostic test that screens for dangerous changes in the cervix, no woman need ever die of this disease. Unfortunately, such tests aren't always available, and cervical cancer is therefore a leading cause of death for women in many developing countries. This is a particular shame, since cervical cancer is also one of very few cancers with a clearly diagnosable and measurable cause—infection with one of the cancer-causing types of human papilloma virus. Improvements in DNA analysis have led to the realization that the vast majority, if not all, cases of cervical cancer are caused by HPV. It can cause other sexually transmitted cancers as well. HPV infection is also associated with rectal, anal, and penile cancer. Hope looms on the horizon, however, since the FDA has recently approved a vaccine to prevent the most common cancer-causing forms of the virus.

CONTAGIOUS CANCER

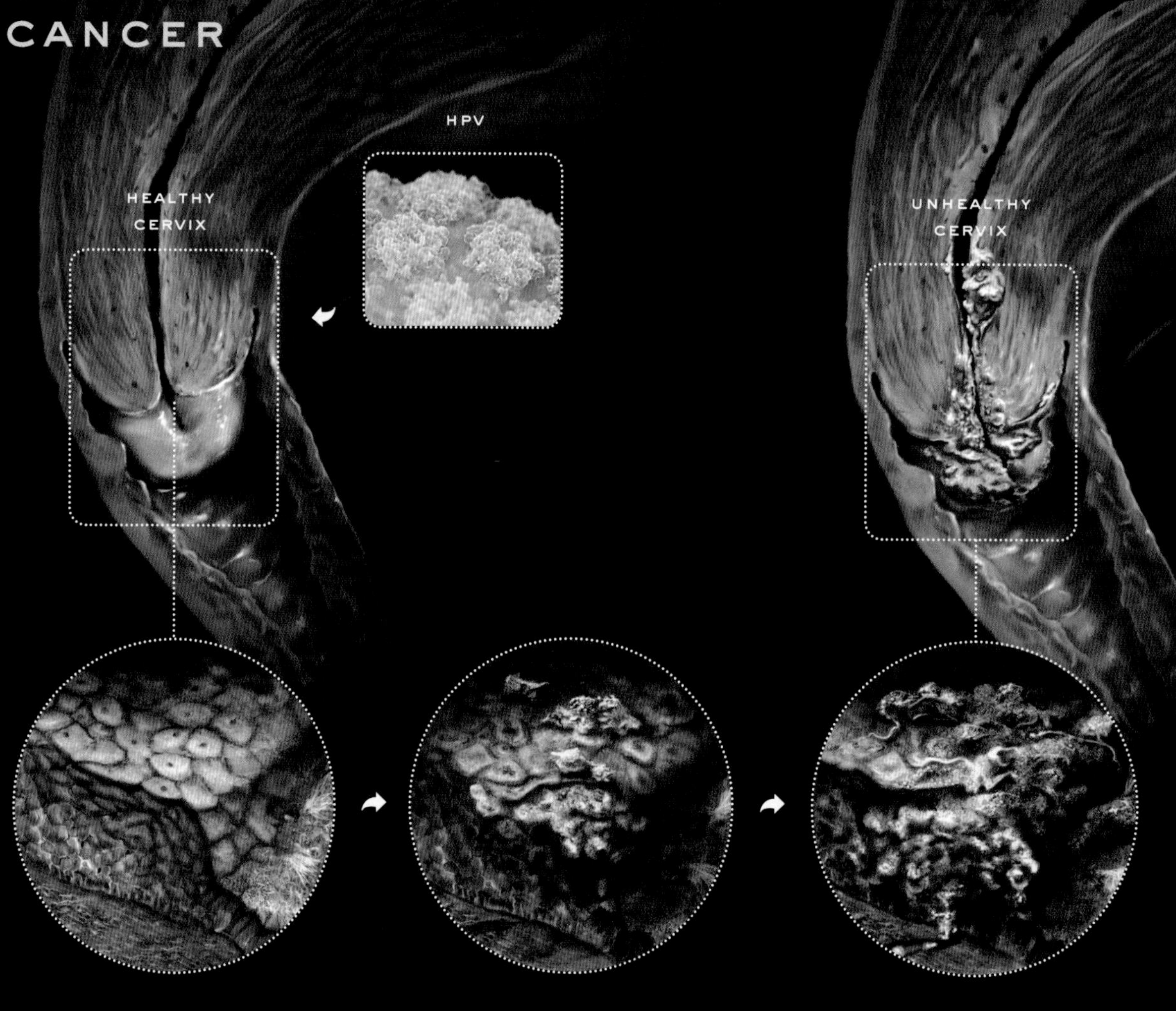

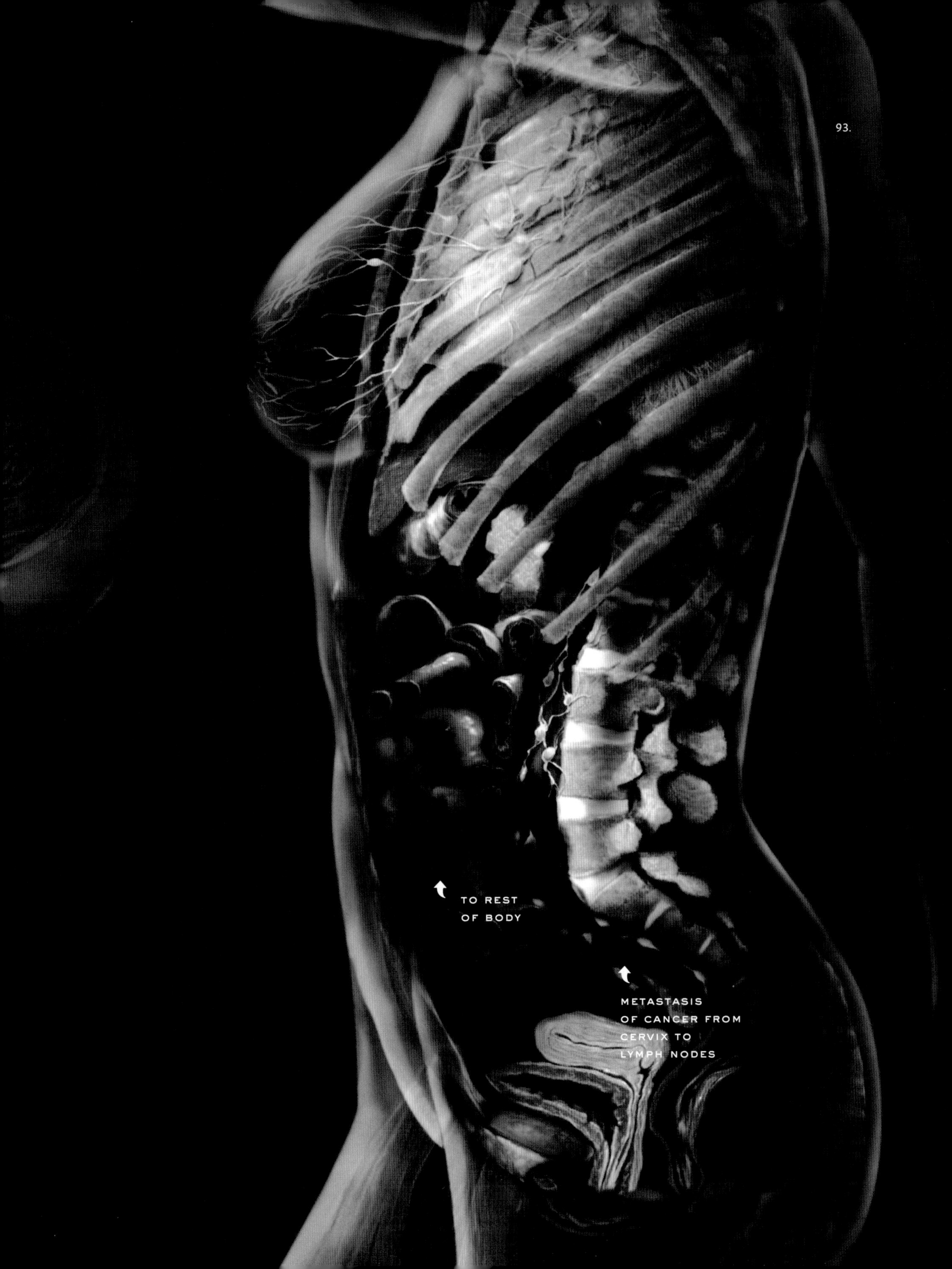
TO REST
OF BODY
METASTASIS
OF CANCER FROM
CERVIX TO
LYMPH NODES

CANCEROUS CELLS

CROSS SECTION OF BREAST SHOWING CANCEROUS TISSUE

CROSS SECTION OF BREAST SHOWING HEALTHY MAMMARY TISSUE AND GLANDS

COPING WITH CANCER

Reproductive cancers are particularly devastating because people sometimes feel that these cancers damage the essence of their masculinity or femininity. Not only can these cancers cause pain during sex, but treatment can disfigure the parts of the body that many associate with sexual attraction. Chemotherapy and radiation for any cancer can cause exhaustion and hair loss, leading to feelings of loss of attractiveness or low self esteem. Losing sexual organs to cancer can be particularly devastating for some individuals. Women may ask, "Will he still love me if I only have one breast?" or "Am I still a woman if I can no longer bear children?" It can be a challenge to focus on the fact that *just being alive and healthy* is beautiful and sexy. We are more than the sum of our parts.

Breast cancer affects approximately one in 93 women by the age of 45, and over the course of a lifetime one in eight women will be diagnosed with the disease. Although the discovery of the breast cancer genes BRCA1 and BRCA2 was hailed as miraculous, in reality, very few cancers are caused by these genes. Women who have these genes, or women who have a family history of breast cancer, are at much higher risk for cancer and should start being screened for it earlier than women who don't. Every woman needs to go for regular mammograms (screening tests) once she reaches the age of 40. All women, even those in their 20s, should perform regular breast self-exams and have a clinical breast exam during their annual medical exam. Early detection is critical, because when breast cancer is diagnosed at an early stage, women have an

UTERINE CANCER

OVARIAN CANCER

excellent chance of survival—greater than 90 percent for stages zero and one combined.

Approximately 1 to 2 percent of women in the U.S. will be diagnosed with endometrial (uterine) cancer. Most endometrial cancer occurs in postmenopausal women, but between 2 to 5 percent of cases occur in women under 40. This condition is different from endometriosis, where the uterine lining grows outside of the uterus. It is characterized by malignant (cancerous) changes in the cells instead of by an abnormality in where they grow. If caught early enough, survival rates are 5 to 9 percent. Fortunately, most cases are diagnosed in an early stage since this cancer has the highly noticeable symptoms of abdominal cramping, heavy vaginal bleeding, and/or bleeding between periods. Although these symptoms are common to many other diseases as well, investigating them usually leads to efficient diagnosis.

Ovarian cancer is uncommon, with a lifetime incidence of only one in fifty women, but it is the fifth leading cause of cancer death in women. Twenty-five percent of ovarian cancer deaths occur in women between the ages of 34 and 55. The disease is so deadly because diagnosis is so difficult. Symptoms are vague, and ovarian cancers shed cells that can invade other internal organs long before the presence of disease is even suspected. More than half of all women with ovarian cancer are diagnosed at an advanced stage, and overall survival rates are only 35 to 38 percent. If the disease is caught early enough, however, the five-year survival rate is greater than 90 percent.

PROSTATE AND TESTICULAR CANCER

Male reproductive cancers hit at the extremes of age. Testicular cancer, although it represents less than one percent of all cancers in men, is the most common cause of cancer in men aged 15 to 40. Most testicular cancers are caught early, diagnosed when a man or his partner notices a change in the shape of his testicles, and the survival rates are excellent. Loss of a single testicle does not affect sexual performance, and an artificial testicle that matches the size and firmness of the original can even be inserted so that sexual appearance is maintained. Loss of two testicles will lead to infertility, but fortunately sperm can be preserved so that if men so choose, they can have children later in life. One of the most interesting things about testicular cancer is that it's one of very few diseases more common in white men than in African Americans.

Prostate cancer is the leading cause of cancer death in men over the age of 75 and the third leading

THE GROWTH OF TUMOR CELLS IN A MAN WITH PROSTATE CANCER CAUSES THE WHOLE ORGAN TO INCREASE IN SIZE.

PROSTATE CANCER

cause of cancer death in men of all ages. It is primarily a disease of older men and almost never occurs in men under the age of 40. It is, in some ways, more difficult to diagnose than other cancers because almost all men experience some enlargement of the prostate as they age. This enlargement is usually benign (noncancerous) and it can be monitored through the use of a blood test for prostate specific antigen (PSA) as well as through more invasive exams. The choice of treatment method can have profound effects on a man's health and sexual function. Early cancers are normally treated with removal of the prostate. Later cancers are often treated by removal of the testicles (termed an orchidectomy, after the Latin word for orchid, whose root the testicle resembles) to lower the level of testosterone and decrease cell growth. Removal of testosterone, however, can have negative effects on both overall health and sexual function.

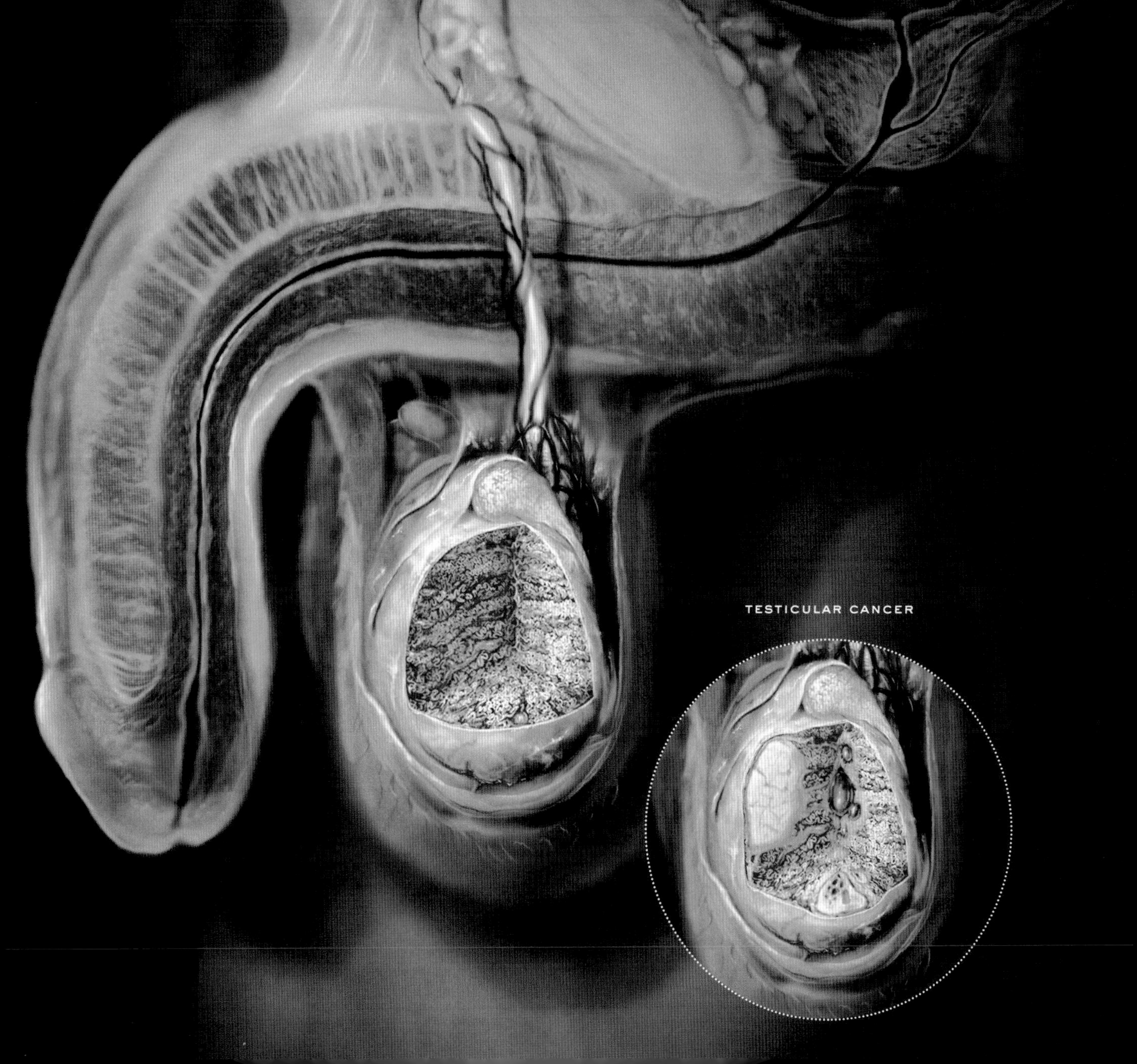

TESTICULAR CANCER

LIFE CAN BE A PAIN

Almost half of all women in the U.S. experience some form of sexual dysfunction. Sexual dysfunction in women can be caused by both psychological and biological problems, and the three most common types of sexual problems are lack of interest in sex, inability to achieve orgasm, and pain during sex. All of these problems are more common in younger women and women with less education, suggesting an important role for psychological maturity and life experience. But a great deal of sexual dysfunction in women is actually biological in origin. Sexual pain, for example, is far more strongly associated with urinary tract symptoms and other health problems than it is with emotional problems or stress, and arousal disorders are equally affected by both.

Vaginismus is an uncontrolled spasming of the outer vagina which can make penetration impossible. It can be caused by physical scarring as well as psychological difficulties, and it is interestingly one of the main causes of unconsummated marriages. Dyspareunia is defined as pain during sex in a man or a woman, not accounted for by vaginismus or lack of lubrication. The most common form of dyspareunia is vulvar vestibulitis syndrome (VVS), where women are extremely sensitive to sensation on the vulva or surrounding glands, and particularly to penetration. Studies have shown that women with VVS can perceive levels of touch that are imperceptible to other women, and that normal levels of touch to most women cause women with VVS extreme pain. Treatment for female sexual pain disorders usually involves psychological counseling as well as techniques that a woman can perform to reduce her sensitivity to penetration. Pain-reducing medications, such as tricyclic antidepressants, may also be prescribed, and some doctors have had success with biofeedback, physical therapy, and other similar treatment methods. There is no one accepted method for treating these disorders, and very few clinical trials have been performed.

Uterine fibroids are benign muscular tumors of the uterine wall. Essentially, the uterus puts out one or more extensions of itself, which can fill the uterine cavity or grow outside the wall into the abdomen. Uterine fibroids can range in size from very tiny (a quarter of an inch) to larger than a cantaloupe. Occasionally they can cause the uterus to grow to the size of a five-month pregnancy. Although they are not inherently dan-gerous, uterine fibroids can cause pain or make it difficult to maintain a pregnancy. Treatment ranges from ignoring them completely to hysterectomy—removal of the uterus.

Endometriosis is a common condition where the endometrium, or uterine lining, grows outside of the uterus. Depending on the extent of the disease, the symptoms of endometriosis can range from none to extreme pelvic pain. There may even be bleeding inside the abdomen, since this tissue experiences the menstrual cycle in the same way as endometrial tissue that is properly confined. More than 97 percent of pelvic pain in women is due to endometriosis. Treatment for endometriosis ranges from use of birth control pills or other hormones to try and control the growth of the tissue to surgery—which may remove either the excess endometrial tissue or the uterus itself.

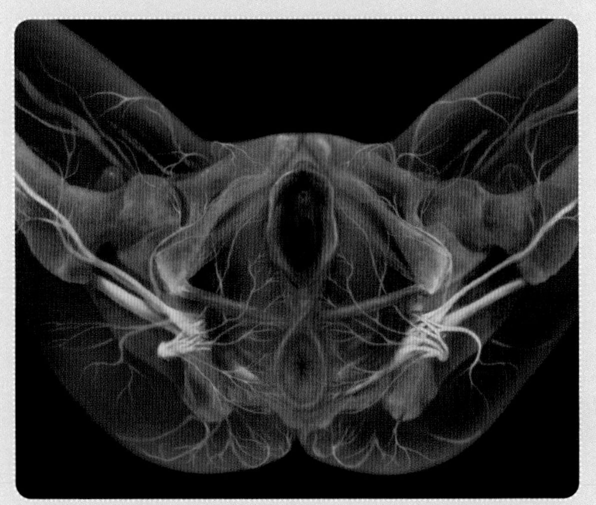

FEMALE PELVIC MUSCLES AND NERVES

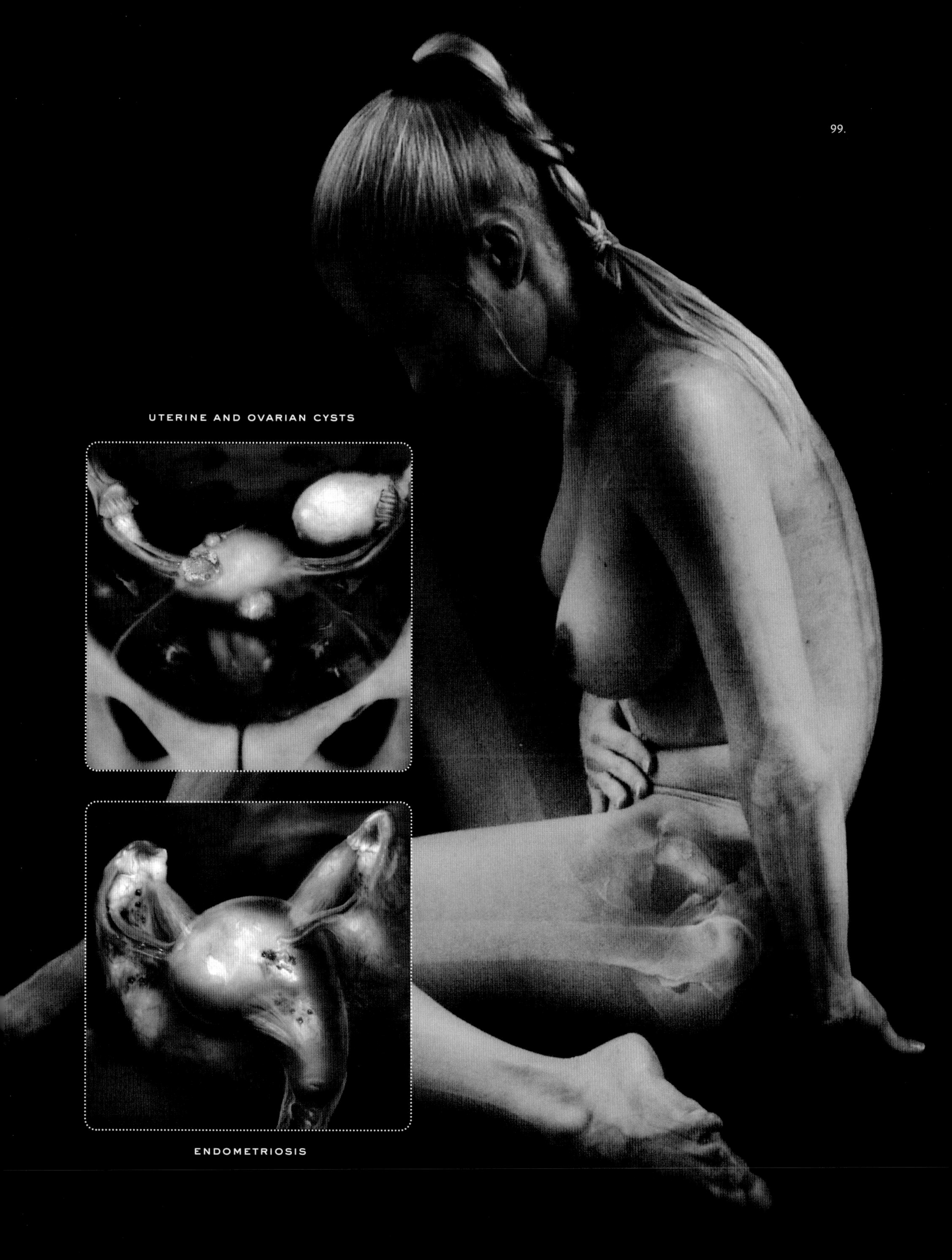
UTERINE AND OVARIAN CYSTS
ENDOMETRIOSIS

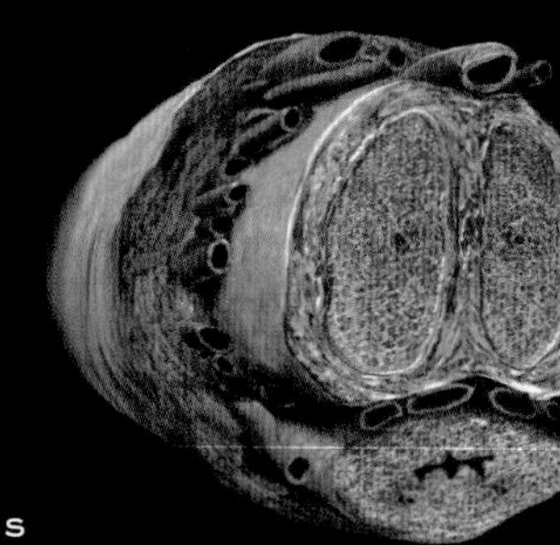

WHEN THE PENIS BECOMES ERECT, IT INCREASES IN FIRMNESS AND DIAMETER. THIS IS DUE TO AN INCREASE IN BLOOD FLOW TO THE ERECTILE TISSUES.

LUMPS AND BUMPS

Priapism is defined as a persistent or undesired erection of the penis, and it is considered to be a problem when regular sexual activity will not make it go down. It's tempting to joke that an endless erection sounds like every man's fantasy, but in fact it's very dangerous. If a man stays hard too long, he can suffer significant medical problems, including permanent damage to the erectile tissue. At some point, usually around the 4-hour mark, medical treatment becomes necessary, which can range from administration of medication to cutting holes in the penis to drain out the excess blood.

Priapism is divided into two types: low flow (ischemic) and high flow (non-ischemic.) In low-flow priapism, which is considered to be the most serious, the veins that allow blood to flow out of the penis are blocked. This is not only a painful experience—if left unrelieved, it can also cause permanent scarring and impotence. One frequent cause is sickle cell anemia, when malformed (shaped like a sickle) blood cells get stuck in the penile veins. It can also be caused by neurological problems, various medications, and certain types of infection. In high-flow priapism, blood simply flows into the penis too fast. This is normally caused by some traumatic damage to the penis, including local injection of drugs, which damage the penile arteries, leading to localized uncontrolled bleeding. It does not usually lead to permanent damage.

For men, there can be multiple sources of pelvic discomfort. Two of the most common ones are varicocele and paratesticular cysts. Varicoceles are essentially varicose veins of the spermatic cord—the structure that supports the testes inside the scrotum. Varicoceles are most commonly found in young men between the ages of 15 and 25. They tend to be located on the left side of the scrotum, probably because of the direction of circulation. They are usually minor and can be managed with the use of a scrotal support. However, if the damage is serious, lack of sufficient blood flow to the testes can result in testicular atrophy and infertility. Microsurgery is an option for the repair of varicoceles.

Cysts of the testes and surrounding tissues come in many forms, and those that occur in the scrotum (but not the actual testes) are almost always benign. Such paratesticular tumors, or cysts, are extremely common—far more so than testicular cancer. Non-cancerous tumors can involve the outer covering of the testes, the epididymis, and even the spermatic cord, and most have no symptoms. Just like testicular cancers, they are normally discovered during a manual self-examination, which men should perform at least once a month, and are very easy to mistake for cancerous lumps. This is why all testicular growths must be carefully examined by a physician, using biopsy, ultrasound, or even MRI. Although most will probably be non-cancerous cysts or other benign intra-scrotal lesions, serious disease should always be ruled out. Cancers are much more curable when caught at an early stage, and so it is essential to seek medical diagnosis for any suspicious lump sooner rather than later.

DAMAGE TO TESTICULAR BLOOD VESSELS CAN LEAD TO DECREASED SPERM PRODUCTION.

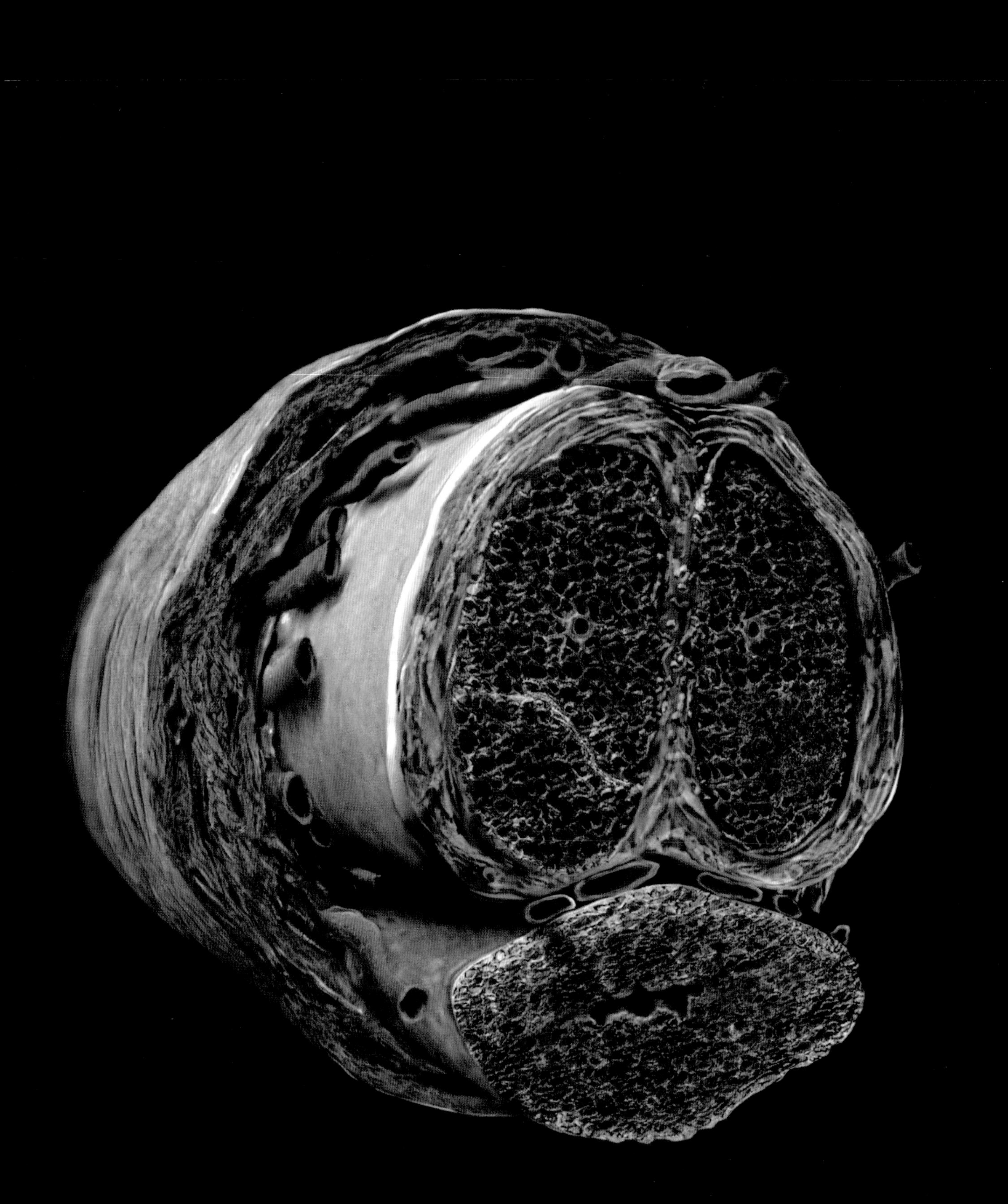

CROSS SECTION OF PENILE ERECTILE TISSUE

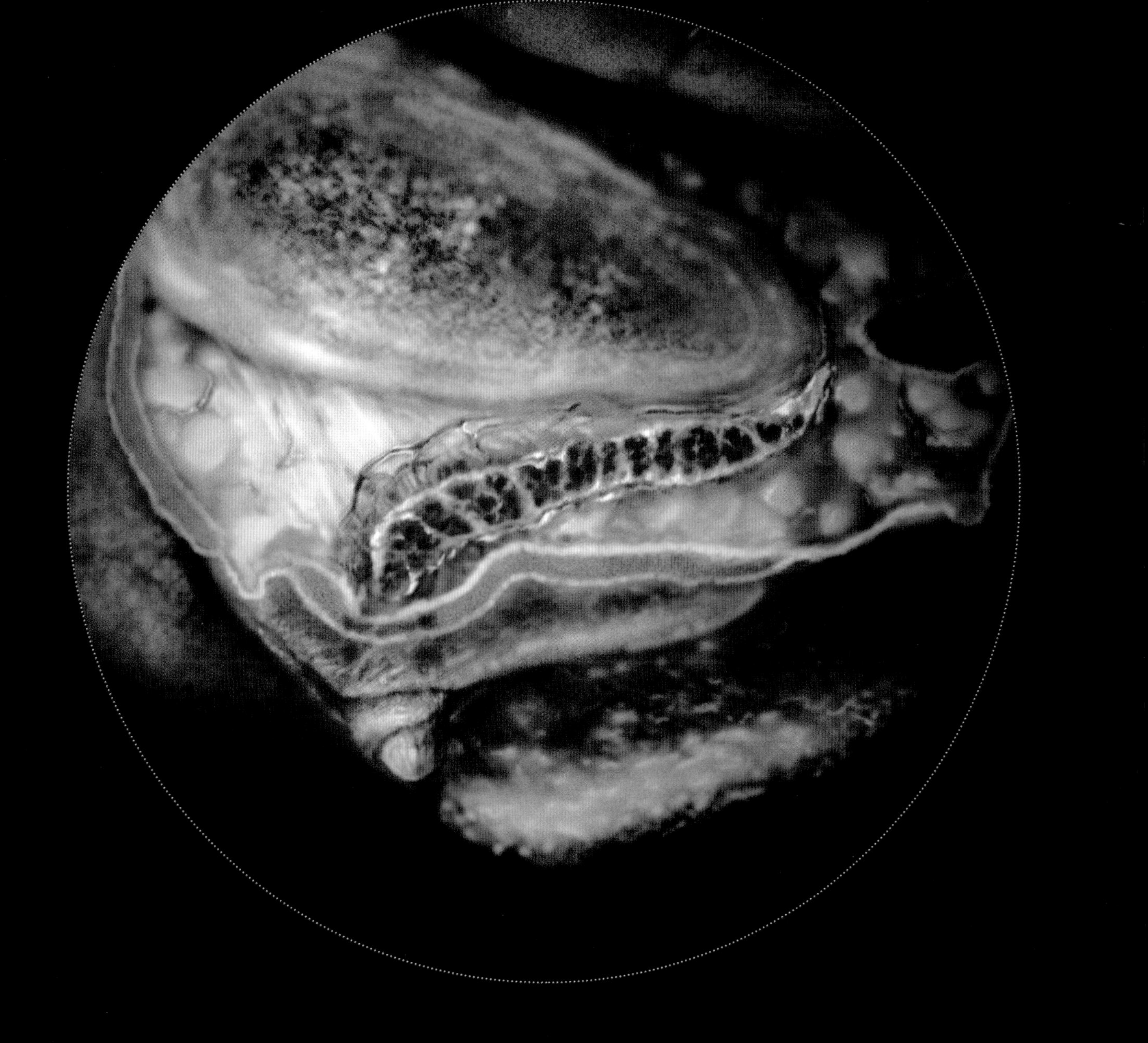

CROSS SECTION OF CLITORAL ERECTILE TISSUE

I didn't lose my virginity until I was 18. The first time was a nightmare. Who shows you how to use a condom?
—ADAM ANT

5

SEXUAL SOLUTIONS

Sex can be enormously satisfying, but perhaps never so much as when it feels like an accomplishment or an adventure. Your first time with a new partner can be traumatic or tremendous, but your first time after an illness is always a triumph. Surviving cancer, sexual difficulties, or reproductive health problems may make you feel like dwelling on what you've lost, but in fact you may have gained much more than you've lost. The changes you've been through, the inner strength you've gained, the appreciation for life itself—all can make sex seem new and exciting again. Treatments for sexual dysfunction can make an enormous difference in the quality of people's lives. Whether it's major surgery, a little blue pill, a change of diet, or even an understanding ear, improving your sexual health can really pay off in enhancing your sex life and your quality of life as a whole.

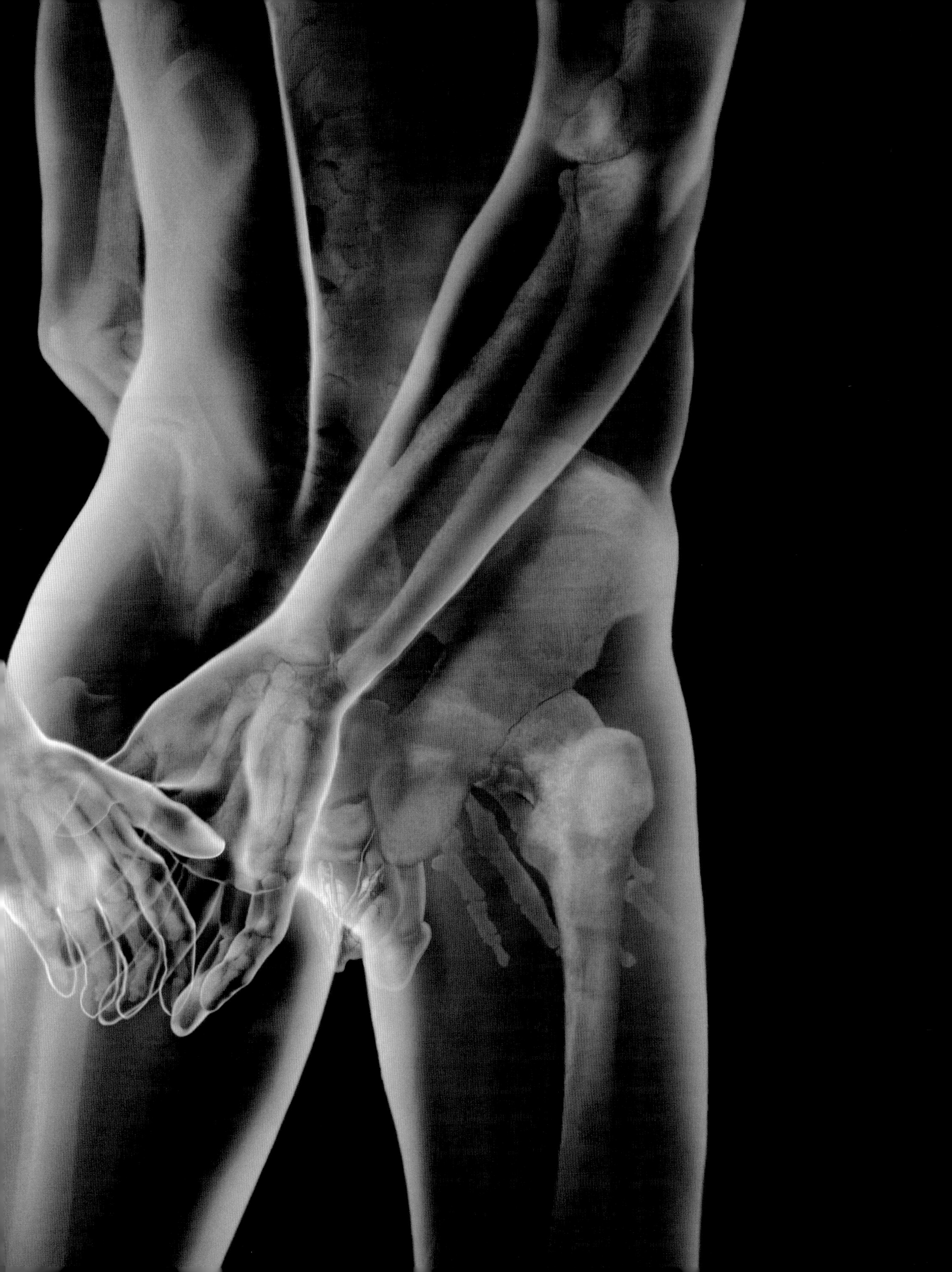

LET'S HAVE A CONVERSATION

CONTRACEPTIVE CONVERSATIONS: Read carefully, all of your options are here.

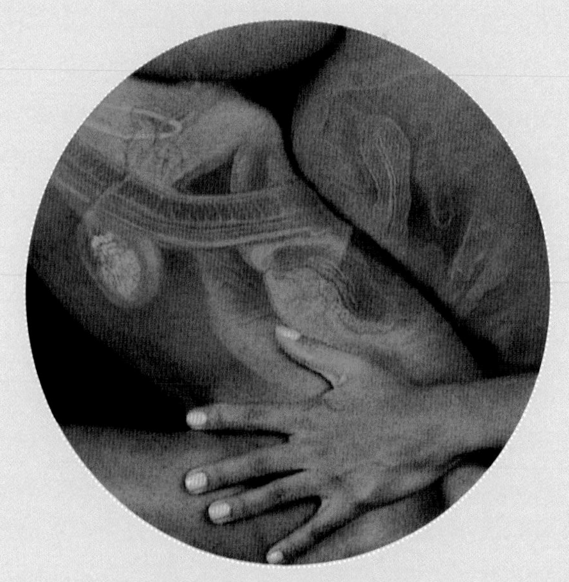

A HANDS-ON CURE: If premature ejaculation is the problem, there's no need to look outside yourself for the cure. You already have the right touch.

THE BIG C: Cancer is a scary word, but if you detect it early enough there is always hope.

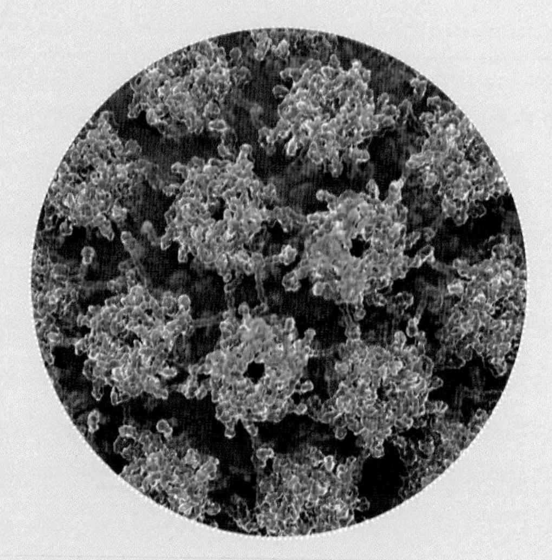

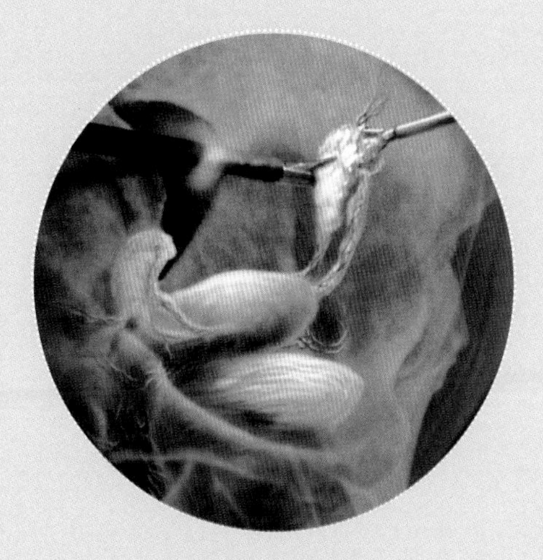

ELIMINATE THE LUMPS: Ovarian cysts are a source of discomfort for many women—but new therapies are constantly being developed.

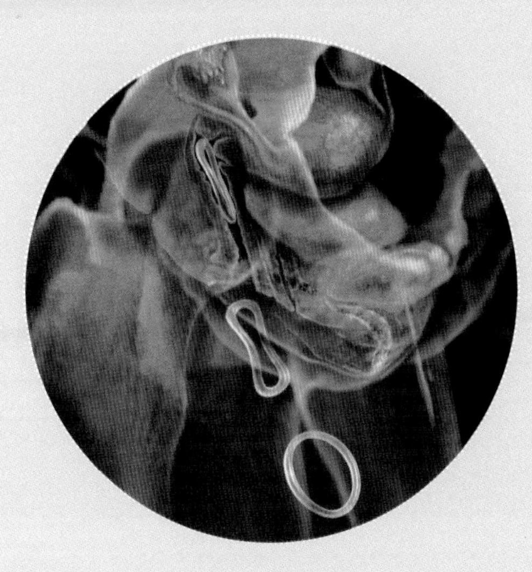

SEX, DRUGS, ROCK AND ROLL?: Are sex hormones the drug of choice for the new millennium?

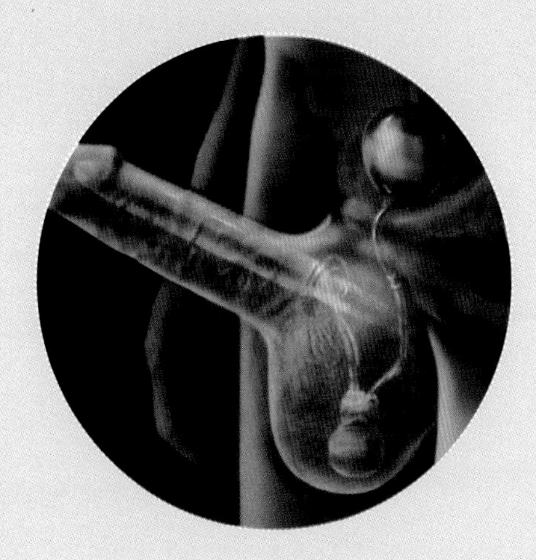

MANDROIDS: Penile implants, male hormonal contraceptives, tiny warheads that fend off sperm and sexually transmitted disease. . . .If these are the future, the future is now.

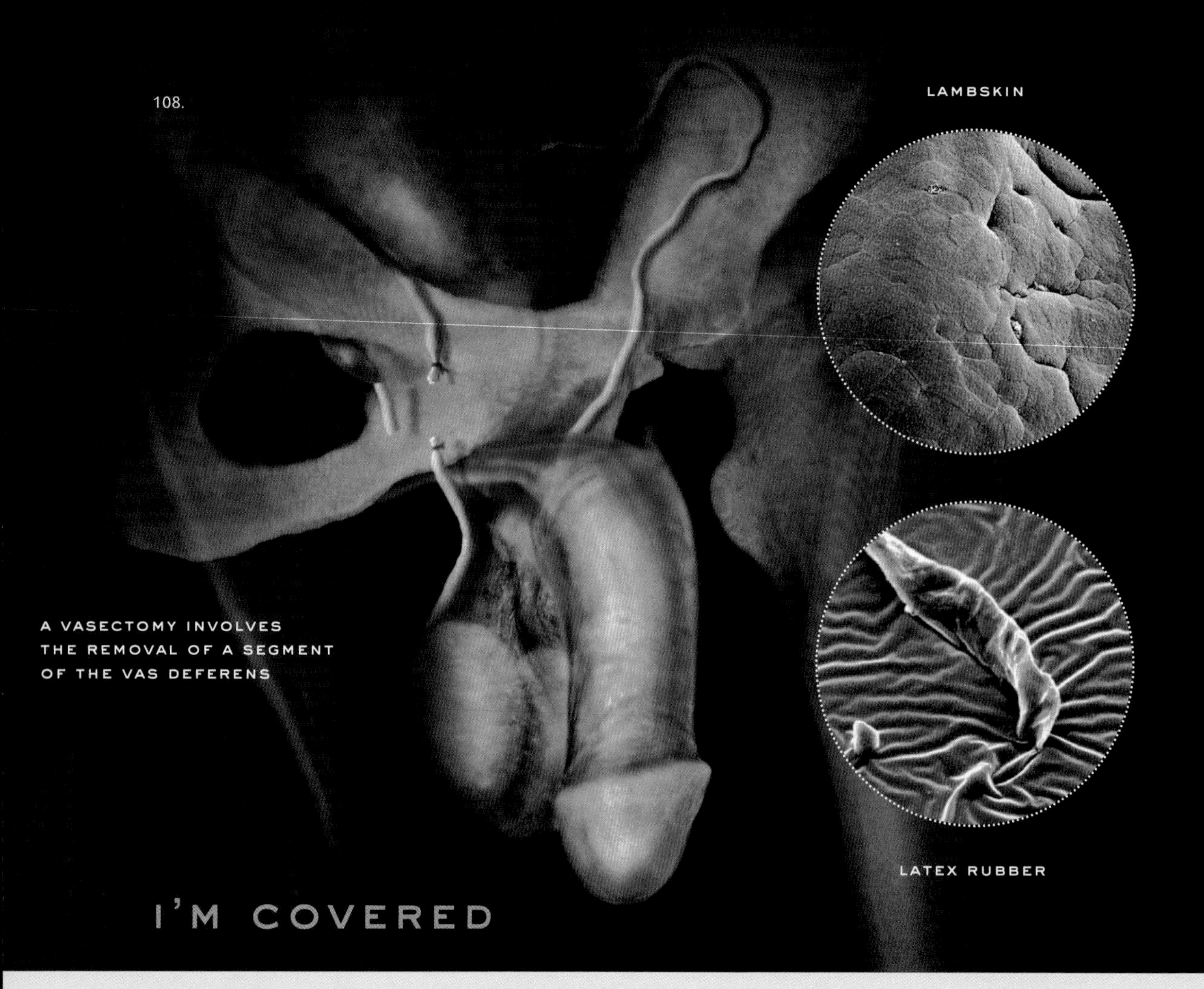

A VASECTOMY INVOLVES THE REMOVAL OF A SEGMENT OF THE VAS DEFERENS

I'M COVERED

It takes two to tango. . .and it takes two to make a baby. Every time a man and woman decide to have vaginal intercourse, there is a chance, so long as the woman is not postmenopausal, that they will conceive a new life.

Contraception is the responsibility of both men and women. But women need to be especially cautious because they bear the brunt of the burden of pregnancy, and because the majority of contraceptive methods are controlled by them. To date, the only effective male-controlled methods of birth control are condoms and vasectomy.

Vasectomy is a type of permanent surgical sterilization that can be performed in less than half an hour, usually as an outpatient procedure. During the procedure, the vas deferens are cut and either clipped, tied back, or cauterized. Healing is generally complete within four to five days. Couples still need to use another form of birth control for two to three months while the sperm still present in the man's body are expelled. For people who follow that instruction, vasectomy is a highly effective procedure with relatively few side effects.

The equivalent procedure in women, tubal ligation, is actually much more invasive. The basic procedure is almost identical; what makes things more complicated is access. Men's vas deferens can be accessed by making a tiny incision in their scrotum, and there are even scalpel-free surgery options available. In contrast, the fallopian tubes are located well inside a woman's abdomen, and even with laparoscopic surgery, accessing them is a riskier procedure.

Condoms fall into the category of barrier methods. They are unique in that they are the only form of

I want to tell you a terrific story about oral contraception. I asked this girl to sleep with me, and she said "No."
—WOODY ALLEN

contraception that also protects against most STDs. Male condoms work by covering the penis and containing the ejaculate. When used correctly, they are the most effective over-the-counter form of birth control available. The important phrase in that sentence is "used correctly." Reservoir-tip condoms were probably invented because too few people knew that they were supposed to leave room at the condom's tip to contain the ejaculate, and it's important to remember that any type of oil—including oil used for lubrication—can cause a condom to break.

There are many types of condoms, but only latex and polyurethane condoms should be used for protection against STDs. The pores in natural skin condoms are so big that viruses can pass right through them, and novelty condoms were never intended as anything other than gag gifts. Although many men have preferences in condom brand, "large" and "extra large" condoms are mostly a matter of vanity. Most standard condoms will stretch easily over a large combat boot, and if the item being covered is bigger than that . . . maybe it shouldn't be going anywhere inside anyone! Not that this is any news to condom manufacturers—at least one company's XXL condoms are exactly the same size as their standard condoms, and that information is clearly visible on the company's Web site. Ironically, some men actually could benefit from petite condoms—condoms do no good if they fall off mid-act—but there doesn't seem to be much market demand for them.

The syphilis epidemic that spread across Europe gave rise to the first published account of the condom, when Gabrielle Fallopius described a sheath of linen he claimed to have invented to protect men against syphilis.[7]

FEMALE PROTECTION METHODS

PRESCRIPTION (RX) REQUIRED?

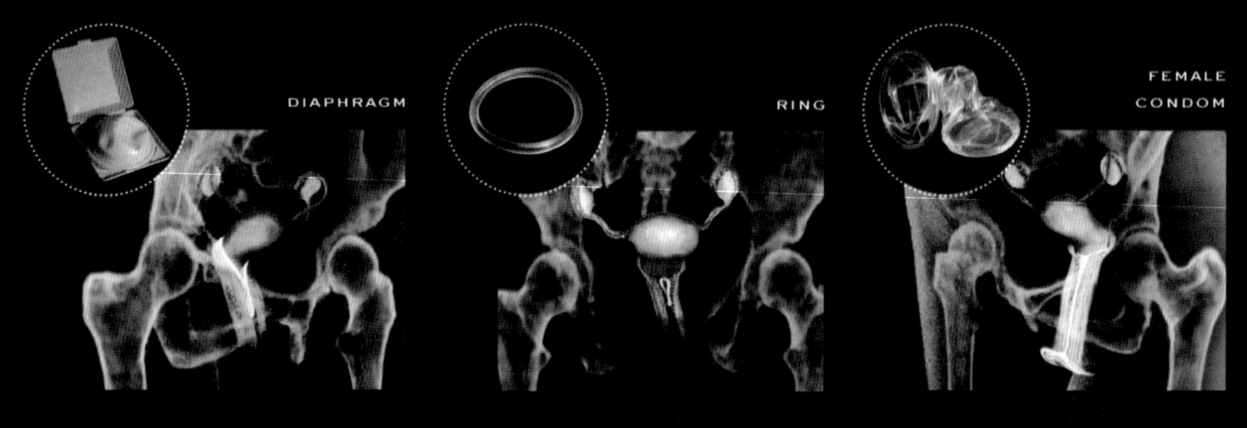

RX: YES

RX: YES

RX: NO

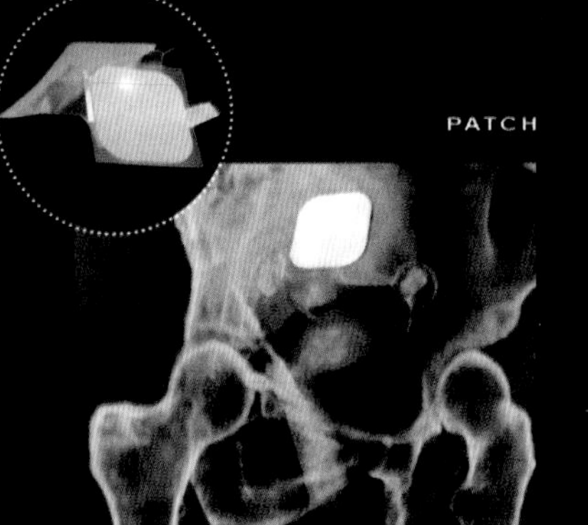

CERVICAL CAP

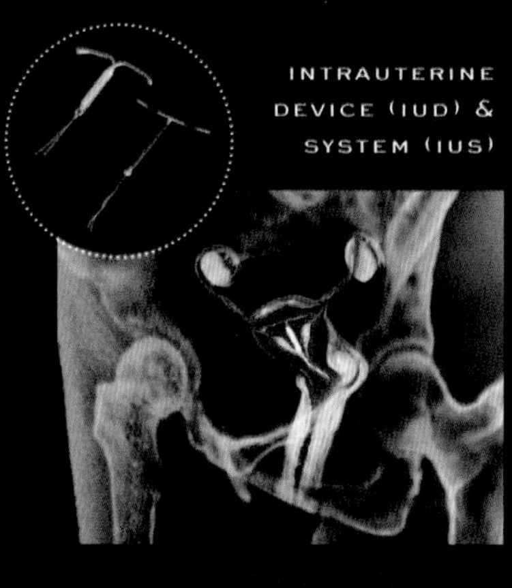

RX: YES

RX: YES

RX: YES

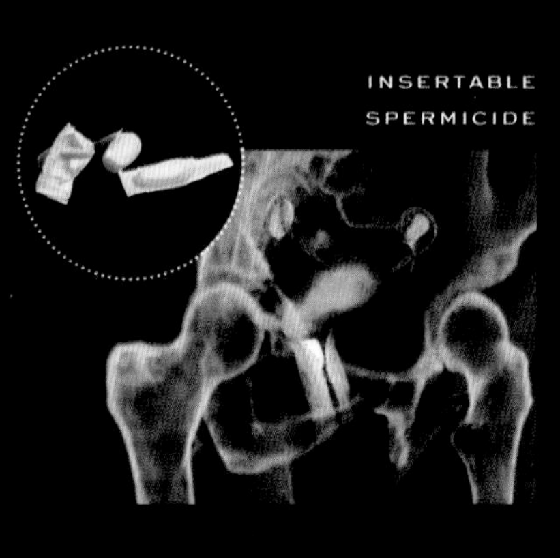

SHOT

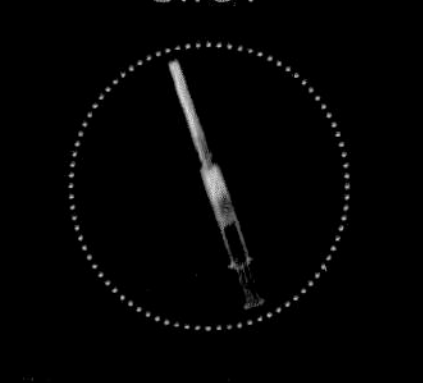

RX: YES

SPONGE

RX: NO

RX: NO

THE PILL

SPERMICIDE

RX: YES

RX: NO

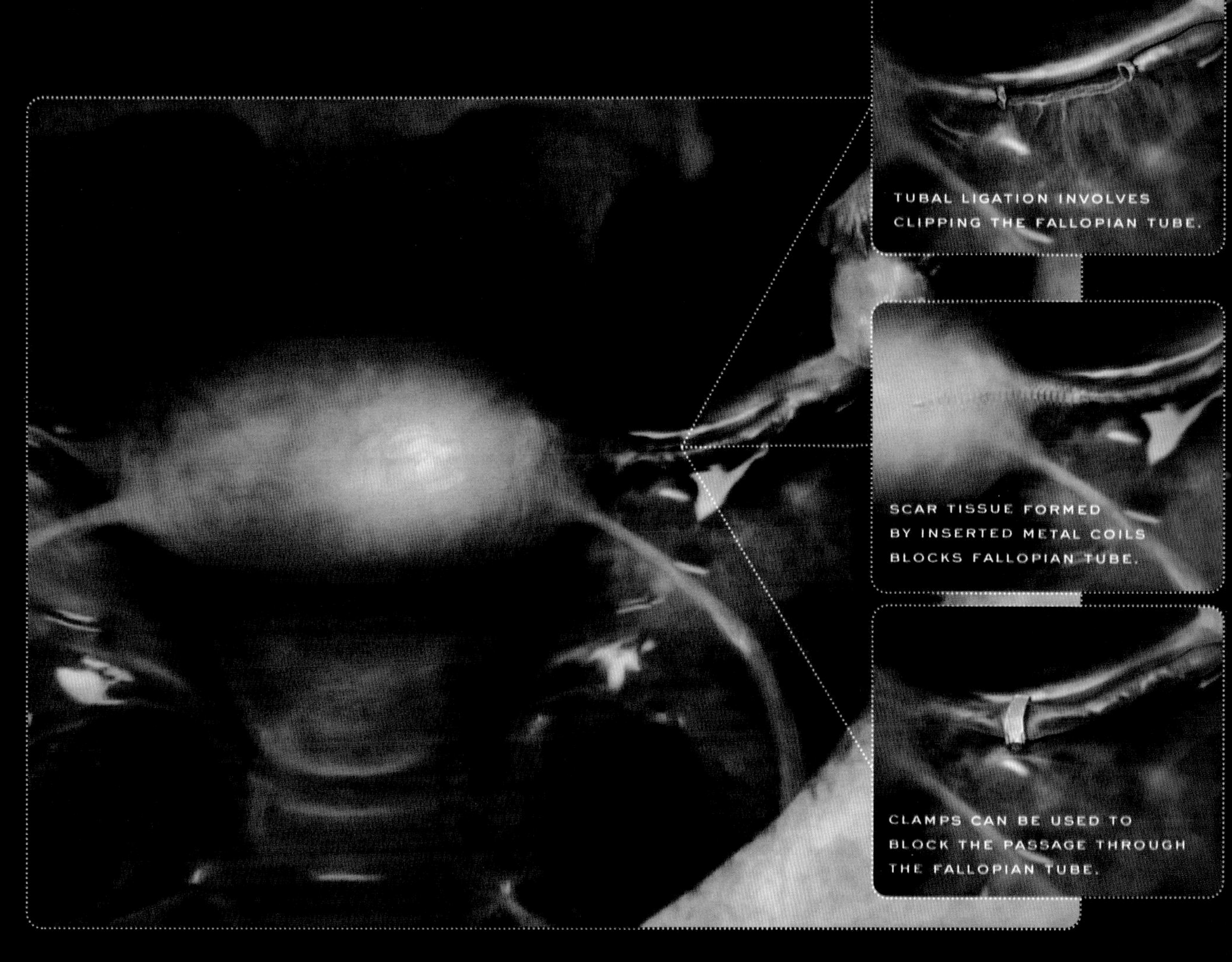

WOMEN HAVE CHOICES

Women have a lot of options when it comes to contraception, but most of their choices are just different sides of the same coin. Female contraceptives are categorized into three main types: hormonal, mechanical, and chemical. Hormonal contraceptives alter a woman's body chemistry so that it either produces no eggs or is otherwise unreceptive to pregnancy. Mechanical contraceptives block the cervix so that sperm can't enter the uterus. Chemical contraceptives kill the sperm before they can cause pregnancy. IUDs, however, don't fit neatly into any of these three categories. They primarily work by making it more difficult for sperm to reach the egg, providing a mechanical barrier as well as thickening the cervical mucus, but they may also mechanically irritate the uterus to make it unreceptive to implantation.

These methods all *can* work, it's just a question of whether or not they *do* work.

Whether a contraceptive works or not has to do with the difference between perfect use and actual use. Perfect use describes how well the contraceptive would work if a woman used it correctly every time she had sex, and ranges from 0.1 percent to 20 percent chance of pregnancy within a year. Actual, or typical, use describes how well the contraceptive works in the real world, and takes into account how well and how often women use the product as well as its basic efficacy. This is the true estimate of contraceptive efficacy, and ranges from 0.1 percent to 40 percent chance of pregnancy in a year.

Premature ejaculation is defined as uncontrolled ejaculation that happens before a man wishes, before or slightly after sexual penetration. It has many causes, including oversensitivity of the penile glans and injury, and it can also be a side effect of some medications. In some men, premature ejaculation is related to psychological issues, and the lack of control over ejaculation can lead to further anxiety—increasing the problem in an unfortunate vicious cycle.

Although in many men premature ejaculation resolves on its own, there are also numerous pharmaceutical treatments for the disorder. SSRIs, a class of drugs used for the treatment of depression, have the side effect of inhibiting orgasms. For men who suffer from premature ejaculation, that's actually considered to be a positive and can lead to off-label use. Inhibition of orgasm allows them to maintain an erection longer than they could without the drugs. PDE 5 inhibitors, which are prescribed for erectile dysfunction, are also useful for premature ejaculation, since they help to prolong erection as well as achieve it. Finally, topical creams and condoms can be used to reduce sensation in individuals whose premature ejaculation is caused by hypersensitivity of the glans.

Behavioral therapy is often the remedy of choice for treatment of premature ejaculation in men who have no obvious underlying cause. The squeeze technique, for example, uses pressure applied to the base of the glans to force blood out of the penis, reduce the size of erection, and delay ejaculation. It can be used multiple times to prolong the sexual experience. Relaxation therapy is also recommended,

HOLD ON TIGHT

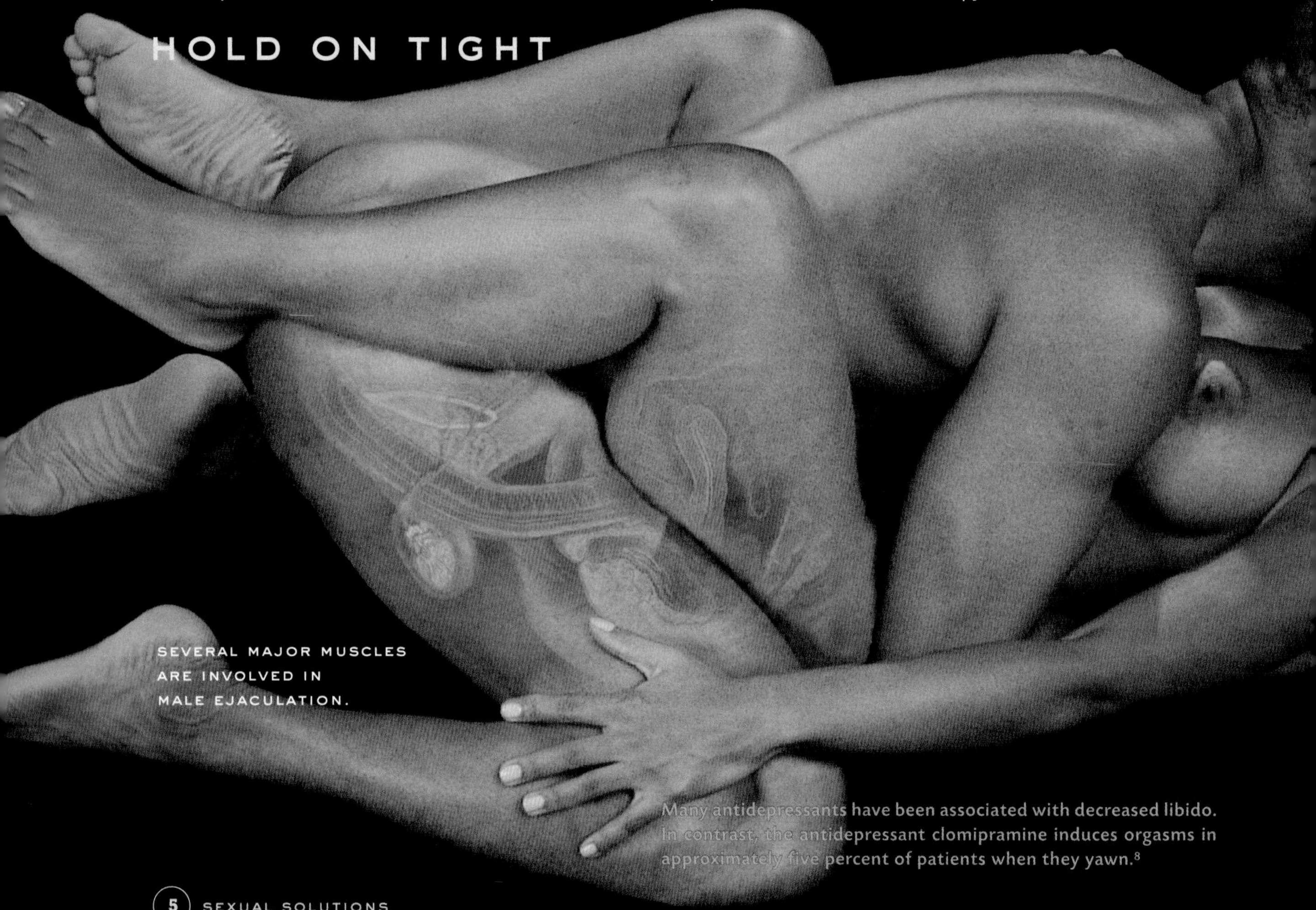

SEVERAL MAJOR MUSCLES ARE INVOLVED IN MALE EJACULATION.

Many antidepressants have been associated with decreased libido. In contrast, the antidepressant clomipramine induces orgasms in approximately five percent of patients when they yawn.[8]

as is paying attention to the specific stimuli that encourage ejaculation and moderating them during the sexual experience.

After the nervous system, the circulatory system plays the most important role in maintaining sexual function. In order for a man to have an erection, blood must pour into the penile arteries and be successfully contained by the penile veins. In order for a woman to lubricate, blood pressure must increase to high-enough levels in her vaginal walls to cause fluid to start to seep through. All of this requires circulatory flexibility, so that blood vessels can dilate and contract on demand and blood pressure can rise and fall as needed.

When individuals have problems with their cardiovascular health, their sexual function can be affected not only by their disease, but by its treatment. Drugs prescribed to control blood pressure may have sexual side effects in both men and women. Simply by successfully fulfilling their purpose and controlling hypertension, they can make it difficult for men to attain an erection or ejaculate. Since vasocongestion is important not only in erection, but in all phases of arousal, tightly controlled blood pressure can also reduce overall sexual response. Unfortunately, since hypertension is mostly without symptoms, sexual side effects can cause people to stop taking their medication and risk their overall health. But there is hope. Certain classes of anti-hypertensive drugs are far more likely to cause sexual dysfunction than others, so people with high blood pressure should be able to work with their doctor to find medication that will preserve both their life and their sex life.

YOUR CARDIOVASCULAR
HEALTH AFFECTS
YOUR SEXUAL WELL-BEING.

TREAT YOURSELF WELL

STDs are commonly grouped into two categories: curable and incurable. The curable STDs are caused by bacteria and protozoa, organisms that can be killed by antibiotics. The incurable STDs are caused by viruses—which are not alive and which can be treated but not destroyed. Antibiotics work in many ways, but they are all selective poisons that are designed to kill bacteria without harming you. They may prevent the bacteria from growing a new cell wall, or stop them from being able to use glucose for energy. However they work, their basic function is to stop the bacteria from reproducing and send them to their doom.

Antibiotic resistance, where bacteria have become immune to the effects of a particular antibiotic, is an increasing problem in this and other countries. Diseases that were once simple to treat now require more and more expensive and complex drugs—if they can be combated at all. There are two simple ways to avoid contributing to this worldwide problem. First, take antibiotics only if they are prescribed by your doctor, and don't pressure your doctor to prescribe antibiotics if you haven't been diagnosed with a bacterial infection. Second, when you are prescribed a course of antibiotics, take every single pill as prescribed, even if you feel better halfway through therapy. If you don't allow the drugs time to kill all the bacteria in your system, the strongest or most adaptable ones will survive. The next time you take the drug, it will be harder for it to kill these bacteria.

Unlike antibiotics, antiviral medications don't work by killing viruses, because viruses aren't alive and cannot be killed. Viruses use a person's own cellular machinery in order to make copies of themselves, so antiviral medications must target the interaction with the host cell somewhere in the infection process. They can prevent the virus from entering the cell, stop its replication, or prevent it from leaving the cell. But because they can't destroy the virus, their effectiveness is usually limited. Antiretroviral therapies for HIV primarily target the replication process, although the newest drug classes also affect cell entry.

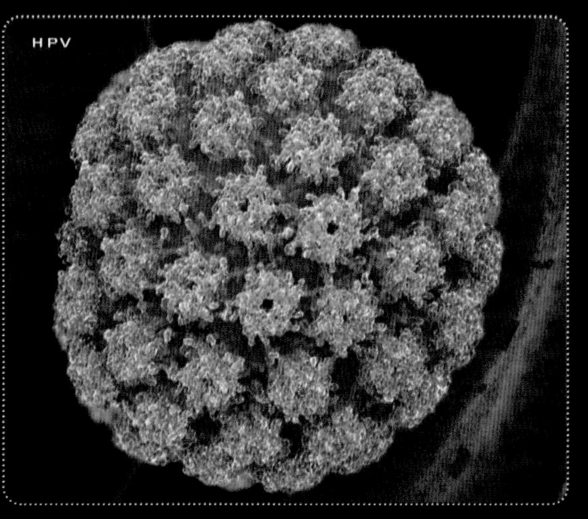

HPV

The newest development in the fight against viral STDs is the HPV vaccine. Vaccines are an artificial way of stimulating your immune system to fight a microorganism that it has not encountered. Vaccines prevent disease by preparing your body to fight in the event that it is exposed to the virus, or bacteria, the vaccine is designed to protect against. STD vaccines are relatively recent innovations. The first such vaccines were against hepatitis; now, vaccines for different types of HPV are in the pipeline. We can only hope that perhaps a herpes or HIV vaccine will be next.

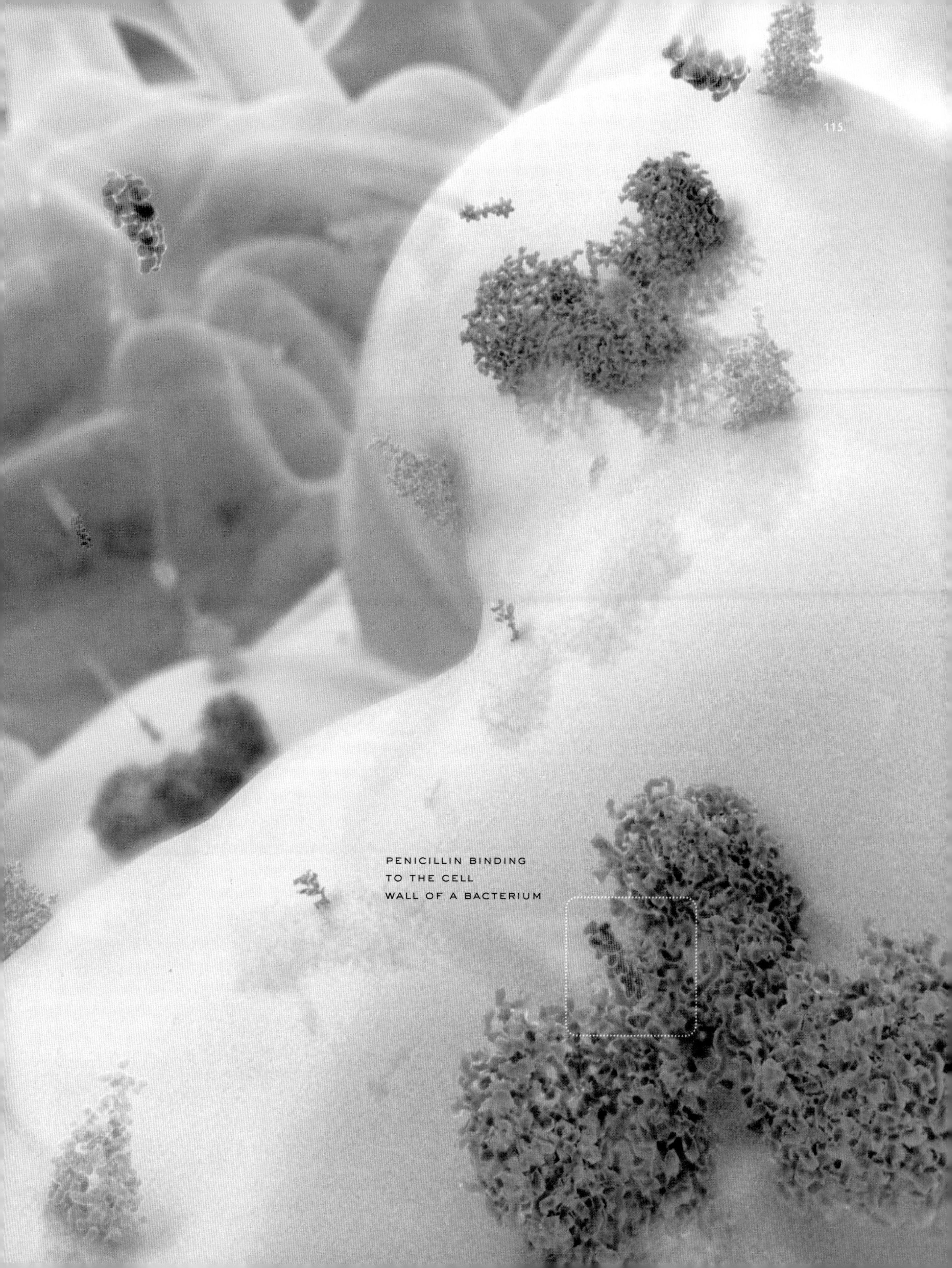

PENICILLIN BINDING
TO THE CELL
WALL OF A BACTERIUM

here are essentially four types of treatment for can-
er: surgery, chemotherapy, radiation, and drugs or vaccines. Which methods are used in any individual lepends on the location of the cancer, how invasive it s, and whether it has spread, or metastasized. Most ancers are treated with a combination of procedures.

Prostate cancer is a difficult cancer to treat, because sometimes it's not easy for physicians to letermine whether it should be treated at all. New echnology has given doctors the ability to diagnose cases of prostate cancer that are not only asymptomatic, but that could possibly have no effect on the patient's health during the remainder of his lifespan. However, it is difficult for many men o understand that not treating their cancer may be he best course of action. Lack of treatment has no sexual side effects, does not require surgery, and vill not affect a man's quality of life. In cases when reatment is necessary, early cancers are usually reated by complete or partial removal of the pro-

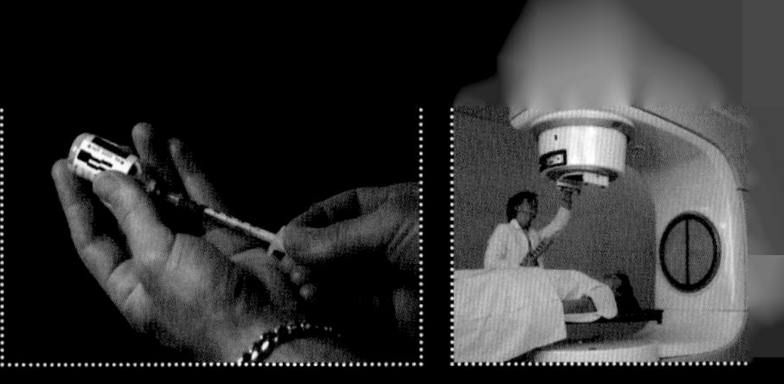

state, and more advanced cancers with removal o the testicles or other testosterone-lowering therapies Both procedures can have serious sexual conse quences, and testosterone replacement therapy may no be an option if there is a chance of cancer recurrence.

Testicular cancer, in contrast, is usually treatec immediately with removal of the affected testicle followed by chemotherapy or radiation. In order tc preserve fertility, a device is used to protect the unaffected testicle from radiation damage. Remova of one or both testicles usually has no effect or sexual function, although occasionally retrograde ejaculation, where sperm are ejaculated backwards into the bladder, may occur. This does not affec

ERASING THE C-WORD

Everybody's favorite question is "How did cancer change you?" The real question is how didn't it change me

—LANCE ARMSTRONG
It's Not About the Bike: My Journey Back to Lif

ORCHIDECTOMY

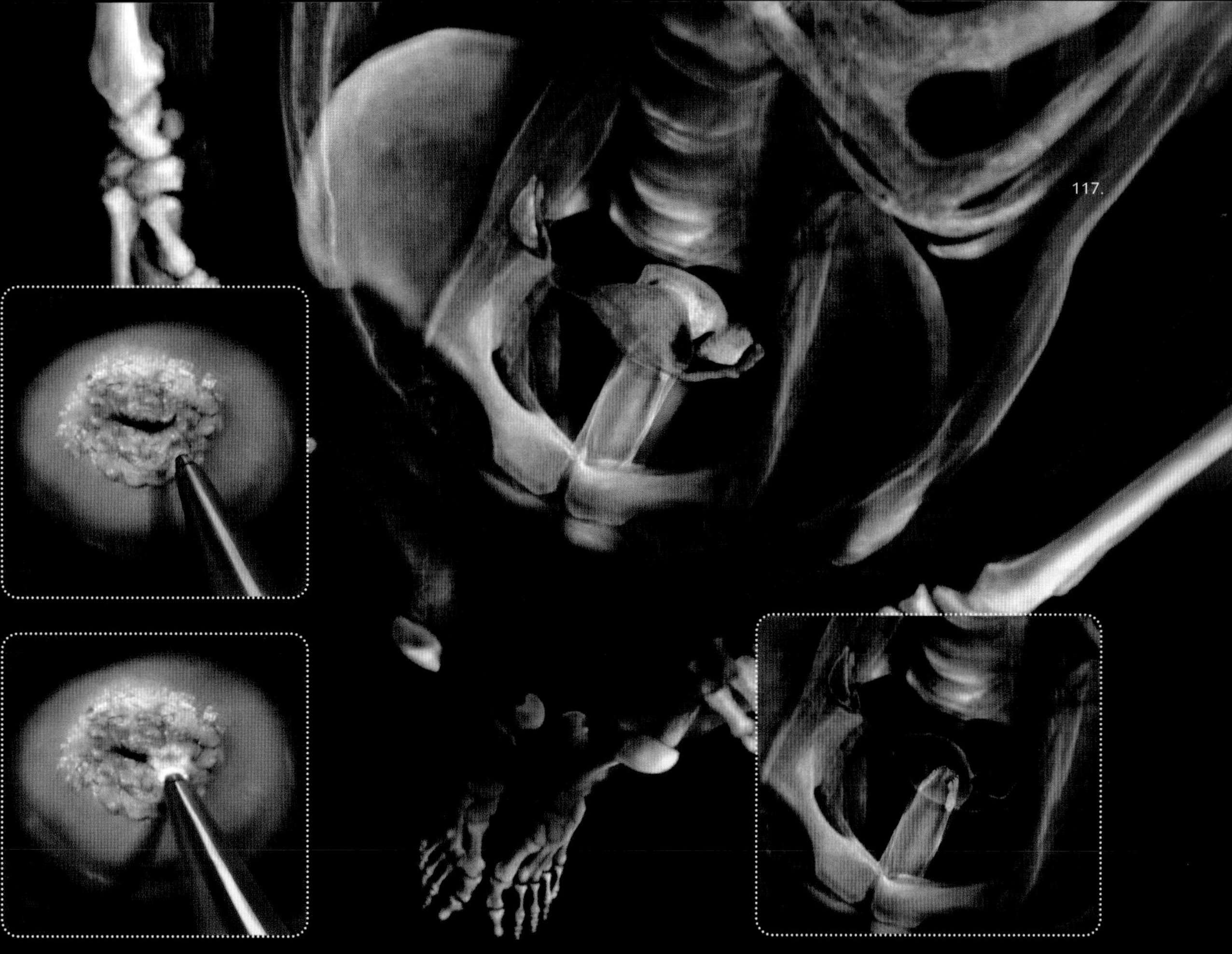

CRYOTHERAPY

HYSTERECTOMY

potency in any way; only ejaculation. Many men choose to have a prosthetic testicle implanted after surgery. Although these devices have no function, they restore the appearance of normality, and can make men more comfortable about resuming their sexual life.

Cancers of the female genital tract have relatively similar treatments, depending on the stage of disease. Ovarian cancer is usually treated with surgical removal of the ovaries, fallopian tubes, uterus, and any sections of affected intestine, as well as intensive chemotherapy. Treatment for endometrial cancer is comparable, but also may include radiation therapy and hormone therapy depending on the nature of the tumor. Cervical cancer, if caught early, does not require such major surgery, and only the affected tissue needs to be removed. However, if allowed to progress to an advanced stage, hysterectomy and chemotherapy or radiation may become necessary. Fortunately, most women with cervical cancer are able to keep their ovaries, which means that hormone therapy isn't necessary to avoid a sudden reduction in levels of hormones, which could cause them to go into early menopause. The benefits of hormone replacement therapy (HRT) are discussed later in this chapter.

Breast cancer treatment is constantly in flux. Whereas for many years mastectomy, removal of the breast, was the standard treatment, more recently lumpectomy along with radiation has been gaining approval. New drugs have also been developed that can bind to hormone receptors in the tumor and actually prevent or arrest the cancerous growth. In stark contrast to the growing trend for less invasive therapies, some women with a family history of breast cancer are actually choosing to have their breasts removed prophylactically—before there is any sign of disease. Finally, although men are far less likely to be diagnosed with breast cancer than women, when they are, their treatment options are essentially the same.

One of the most common treatments for a benign testicular cyst is . . .no treatment at all. Due to recent advances in ultrasound technology, it is possible to easily monitor cysts for any changes that could lead to cancer formation, making it unnecessary to attempt removal if the cyst isn't causing any symptoms. Other treatments for testicular cysts range from excision of the cyst to removal of the entire affected testicle. The last, known as orchidectomy, was once the most frequent treatment for testicular lesions, since it was straightforward to perform and removed the risk of the cyst recurring or becoming cancerous.

One of the major concerns in the diagnosis and treatment of ovarian cysts is the possibility that they may be cancerous. Ovarian cancer is rare, but deadly, and doctors want to make certain not to miss any chance for diagnosis. Fortunately, most cysts are non-cancerous, and they can be examined using sonography—a non-invasive procedure that uses sound waves to see inside the abdomen. Benign cysts in premenopausal women who are still having their periods usually don't even need to be treated, since they are a natural, if unpleasant, consequence of ovulation. However, if the cysts become painful, or continue to grow over a period of several months, they can be removed by surgery. Surgery is also almost always recommended for women who are no longer menstruating, since they should not

SMOOTHING OUT THE LUMPS

MICROSURGICAL TECHNIQUES CAN BE USED TO REPAIR THE DAMAGED BLOOD VESSELS IN A MAN WITH A VARICOCELE OR TO DRAIN A TESTICULAR CYST.

be forming benign cysts. Removal allows for biopsy of the cyst and removal of the ovary in case it might be cancerous. Today, many cysts don't need to be treated at all, because they can be prevented. In young women prone to ovarian cysts, oral contraceptives are used to stop their formation by preventing ovulation.

Until relatively recently, the only two treatment options for uterine fibroids were surgical: hysterectomy—removal of the uterus, and myomectomy—excision of the fibroid tumors. For women who did not need to preserve their fertility, hysterectomy was usually the recommendation, as it was an easier procedure to perform and there is no risk of recurrence. Fortunately, a new procedure was developed for the treatment of fibroids that does not require surgery: uterine artery embolization (UAE). During this procedure, small particles are injected into the arteries supplying the fibroid tumors. This blocks their blood supply and causes them to eventually regress into the uterine wall. It's still a relatively new technique, but data seems to suggest that it's as safe and effective as hysterectomy while, theoretically at least, preserving a woman's ability to bear children. As time goes on, and women having fertility problems because of fibroids are treated with UAE, scientists will begin to develop a clearer idea of this technique's place in the treatment lexicon.

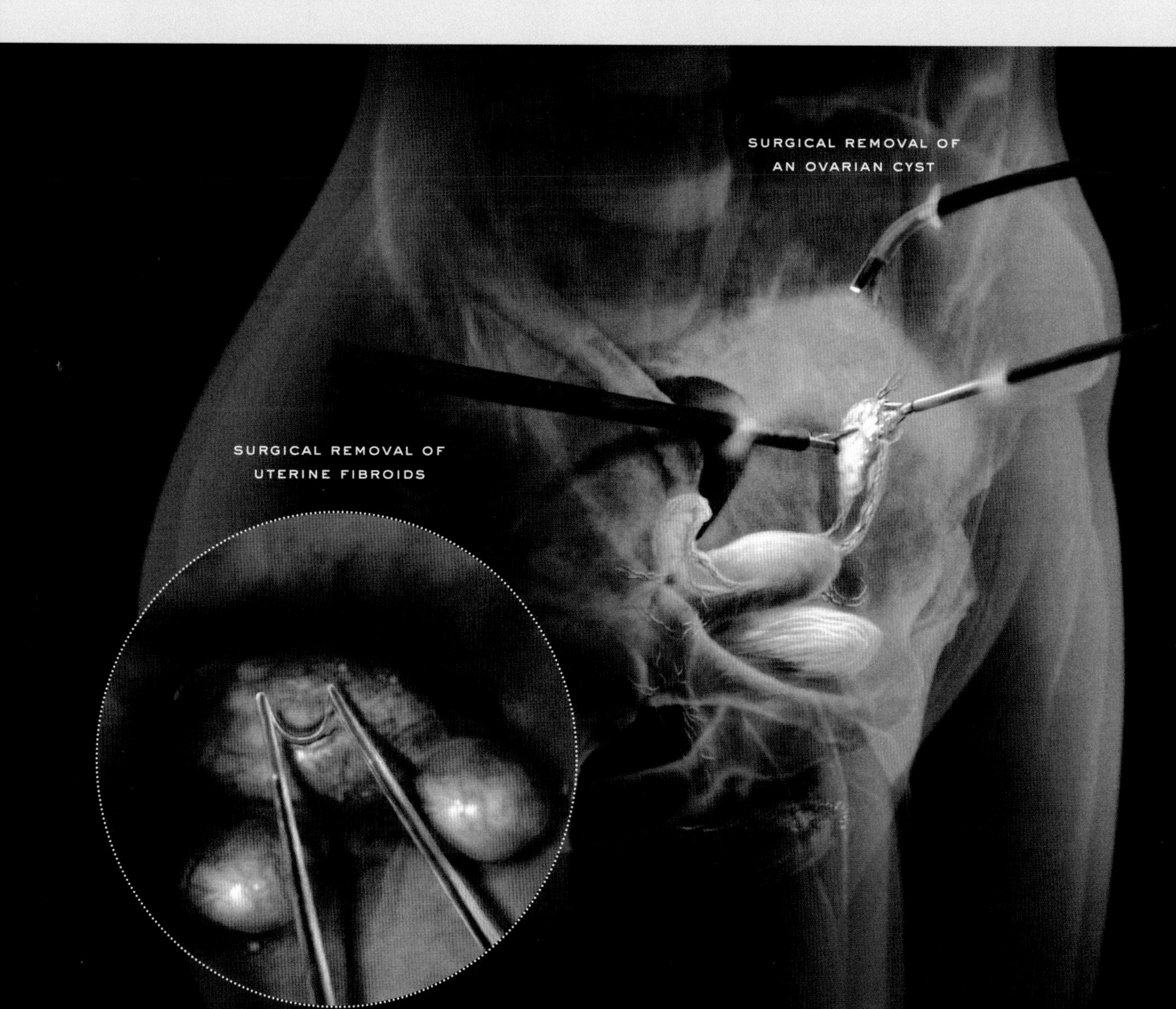

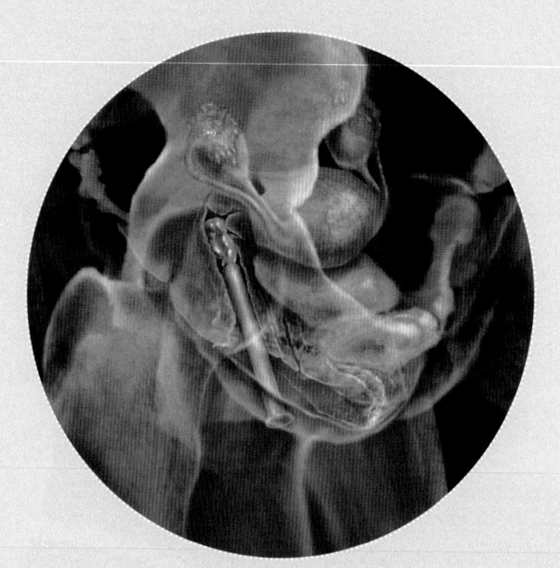

HORMONE REPLACEMENT THERAPY:
INTRAVAGINAL CREAM

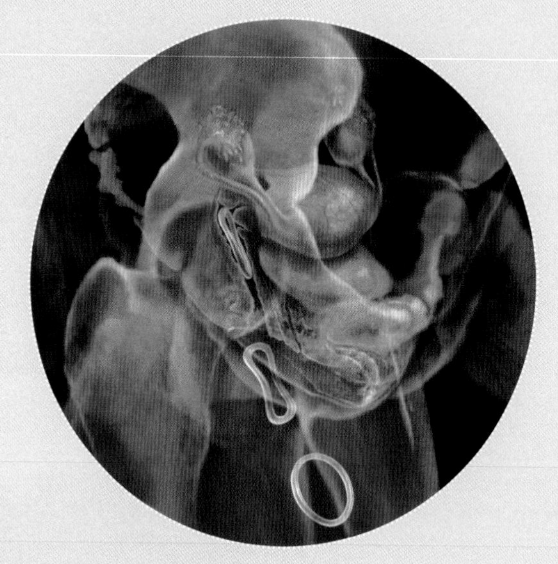

HORMONE REPLACEMENT THERAPY:
VAGINAL RING

CHANGE OF LIFE

Although the process of menopause is universal, each woman's experience of it is unique, and there may be a cultural component as well. Western women have far more negative experiences of "the change" than their Eastern counterparts, and debate has long raged over how much of the difference is cultural and how much biological. Over the past ten years, however, an increasing amount of research has suggested that a diet rich in consumption of soy, with its phytoestrogenic (plant-estrogen) compounds, can actually ease the transition into menopause.

Estrogen replacement therapy in menopausal women, despite the controversy over its general health effects, does lead to definite improvements in sexual response, particularly by increasing vaginal lubrication and wall thickness. It is available in many forms, from pills and injections to skin patches and vaginal suppositories, gels, and rings. Some of these treatment regimens increase hormone levels throughout the body. Others have a more localized response that can be an advantage for women who are worried about systemic side effects or increase in cancer risk. In general, the recommendations for hormone replacement therapy in men and women are similar. Use them, doctors say, if the symptoms require it, but if you don't need them, it's best to let your body have its way. Many individuals maintain healthy sexual function long after the onset of menopause, but for those who don't, hormone replacement therapy can help to improve the quality of their sexual life.

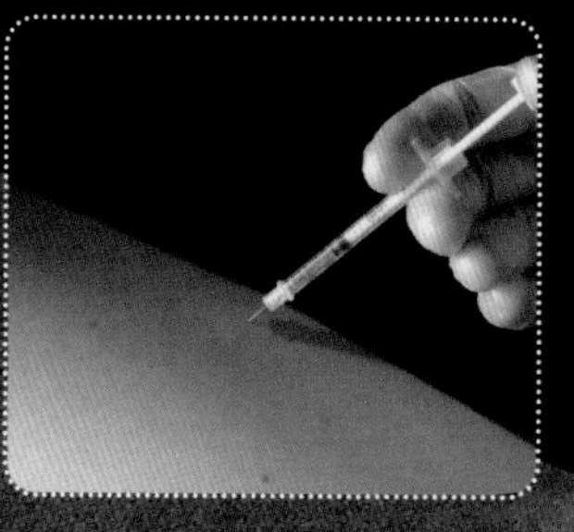

HORMONES CAN BE TAKEN BY INJECTION FOR A SYSTEMWIDE RESPONSE.

TOPICAL CREAMS AND GELS CAN BE VERY EFFECTIVE METHODS OF HORMONE DELIVERY AND ARE ONLY APPLIED WHERE NEEDED.

TRANSDERMAL PATCH

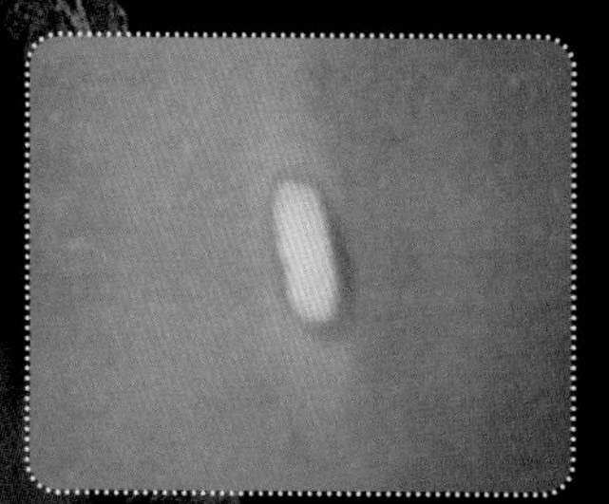

SUBDERMAL IMPLANT

PILLS DELIVER HORMONES SYSTEMICALLY FOR WOMEN WHO MAY PREFER THEM TO INJECTIONS.

Hormones are the building blocks of sexual function. When levels of hormones are low, sexual desire may be absent, and sexual arousal may be difficult or impossible to experience. Although for many years decreasing levels of sex hormones were thought to be a natural part of aging, recently more and more scientists have accepted that the changes in lifestyle that follow their loss can be not only disheartening, but dangerous. For many individuals, the lack of an enjoyable sex life can lead to depression and an overall decrease in physical well-being.

Testosterone therapy helps to maintain sexual interest and function in both men and women. In addition, it appears to help both mood and body composition in men, although as with estrogen replacement therapy in women, the non-sexual benefits of testosterone replacement therapy are still controversial, since there is no detailed understanding of its risks when used in men with low levels of testosterone.

There is one use of testosterone where the risks are well understood. Most men think of steroids as something that body builders take to help them develop big muscles. What they forget is that those drugs are mostly synthetic forms of testosterone. It's the well-known side effects of steroid use in young men that have made many physicians leery of prescribing testosterone replacement therapy in men with low androgen production.

Such a worry is, however, probably unfounded. The goal of testosterone replacement therapy is to maintain hormones at a healthy level, not pump them up to achieve unrealistic goals. A healthy amount of testosterone acts almost as a fountain of youth, improving not only sexual function, but overall fitness levels and quality of life.

VIM AND VIGOR

WHEN TAKEN IN PILL FORM, HORMONES ARE ABSORBED THROUGH THE LINING OF THE SMALL INTESTINES INTO THE BLOOD STREAM.

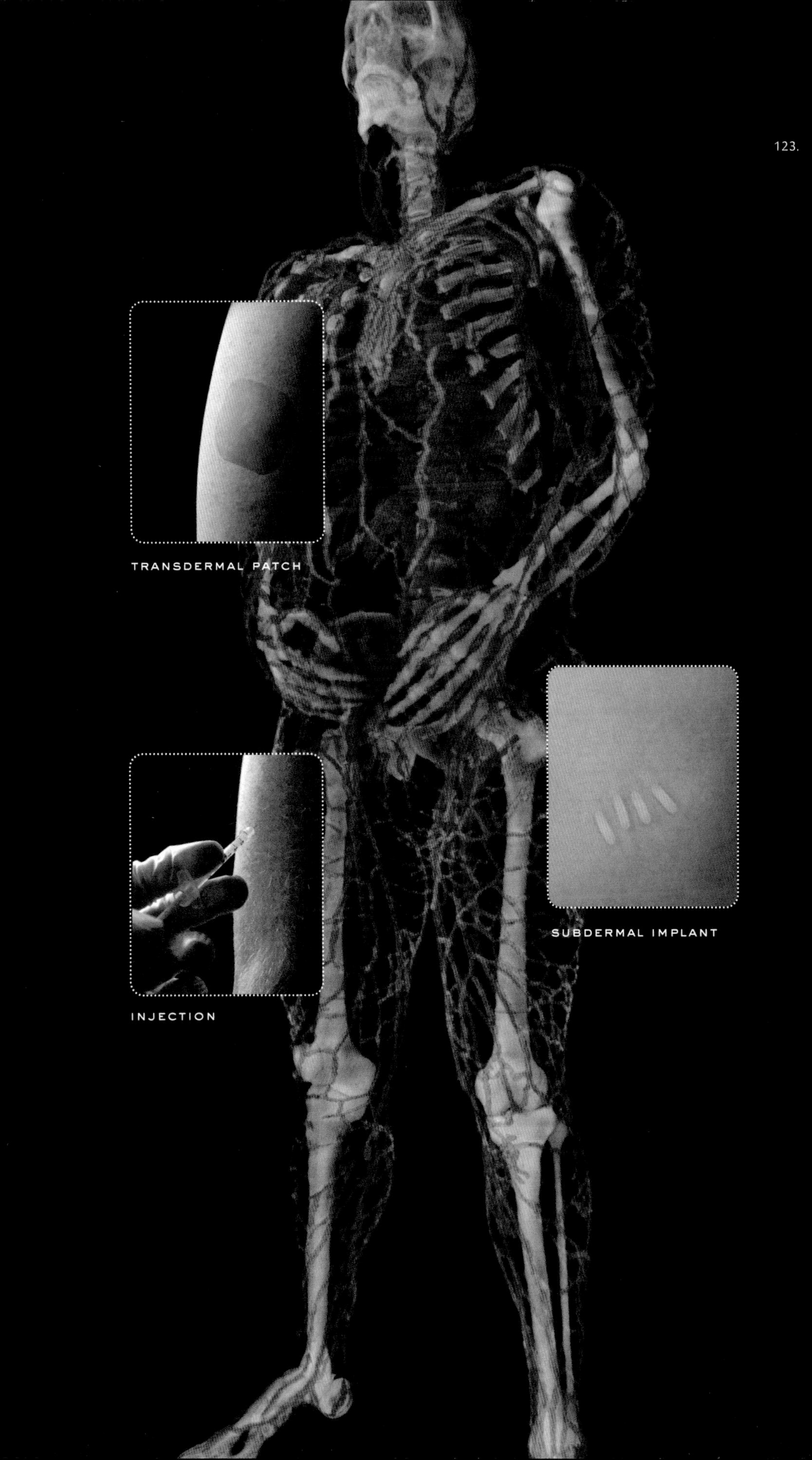
TRANSDERMAL PATCH
SUBDERMAL IMPLANT
INJECTION

For individuals with sexual disorders who either aren't responsive to or don't want to pursue pharmaceutical treatment, it may be possible to use other methods to increase blood flow to the genital region. There are currently vacuum devices available which stimulate the penis or clitoris by, essentially, sucking enough blood into the area to cause an erection. In women, clitoral therapy devices have been shown to improve all facets of arousal, from desire to sensitivity and lubrication. In men, vacuum therapy has been a long-accepted treatment for erectile dysfunction.

Where the presence of a vacuum device is too intrusive during the sexual experience, researchers have explored the use of vasodilating creams and gels to achieve the same effect. These solutions contain compounds, such as the capsaicins found in hot peppers, which increase blood flow to the areas where they are applied. In both men and women, such topical agents have been shown to promote arousal in preliminary studies, and in the future these creams might become an effective, minimally invasive form of treatment for sexual dysfunction.

OTHER OPTIONS

CRURAL LIGATION SURGERY IS PERFORMED TO REDUCE THE AMOUNT OF BLOOD LEAKING OUT OF THE PENILE VEINS

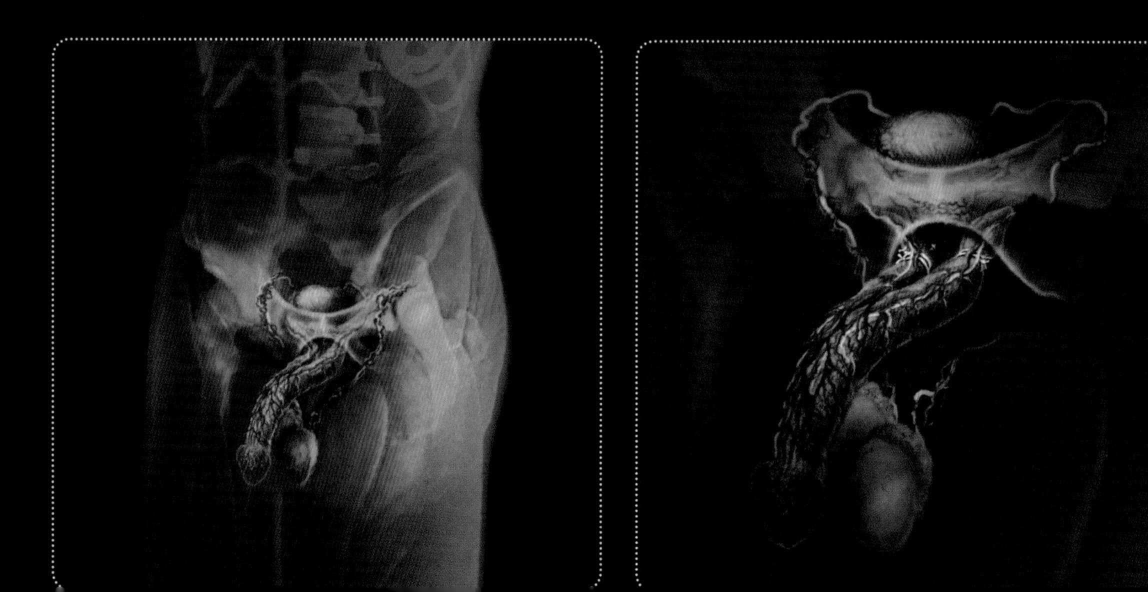

In men for whom medication or external devices are not effective, there are also numerous surgical treatment options for erectile dysfunction. Penile prostheses, for example, can be an effective and permanent way of treating erectile dysfunction. These devices come in two types, semi-rigid and inflatable. Inflatable implants work by allowing a man to manually pump fluid into tubes inserted along the corpus cavernosa. Different designs either expand only the girth of the penis, or both girth and length, and they require sufficient manual dexterity and flexibility to operate the pump. Semi-rigid implants are either rod-shaped or flexible like a gooseneck lamp. They are less expensive than inflatable implants and require less expensive surgery; however, they are less adjustable once inserted.

Other types of surgery for erectile dysfunction generally focus on improving or restoring blood flow within the penis. Crural ligation, for example, helps to reduce the amount of blood leakage from the penile veins, which, when present in abnormal amounts, can make it impossible to achieve or maintain erection. In contrast, penile revascularization improves the amount of blood flow into the penis, providing sufficient blood volume for erectile function in men with damaged or insufficient arteries.

INFLATABLE IMPLANTS ARE A SURGICAL OPTION THAT CAN HELP A MAN WITH ERECTILE DYSFUNCTION MAINTAIN AN ERECTION.

Alas, the penis is such a ridiculous petitioner. It is so unreliable, though everything depends on it—the world is balanced on it like a ball on a seal's nose.

—WILLIAM GASS,
Metaphor and Measurement

MONOCLONAL ANTIBODIES COULD TARGET SPERM CELLS TO PREVENT THEM FROM FERTILIZING THE OVUM.

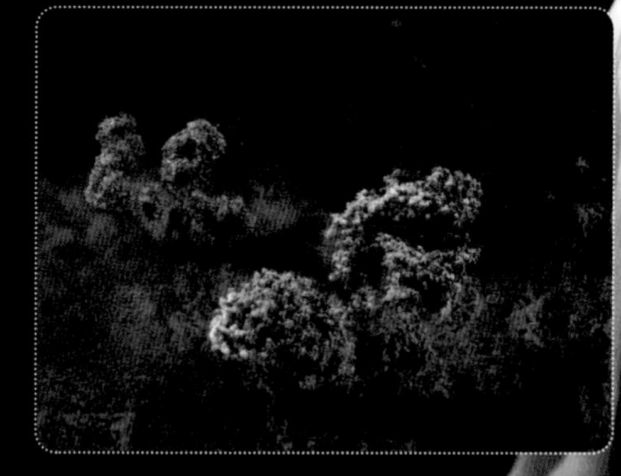

GENE THERAPY COULD BE USED TO RENDER THE SPERM INCAPABLE OF PENETRATING THE EGG.

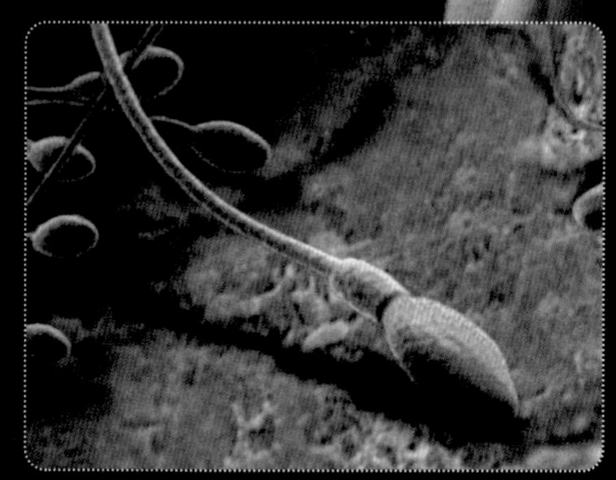

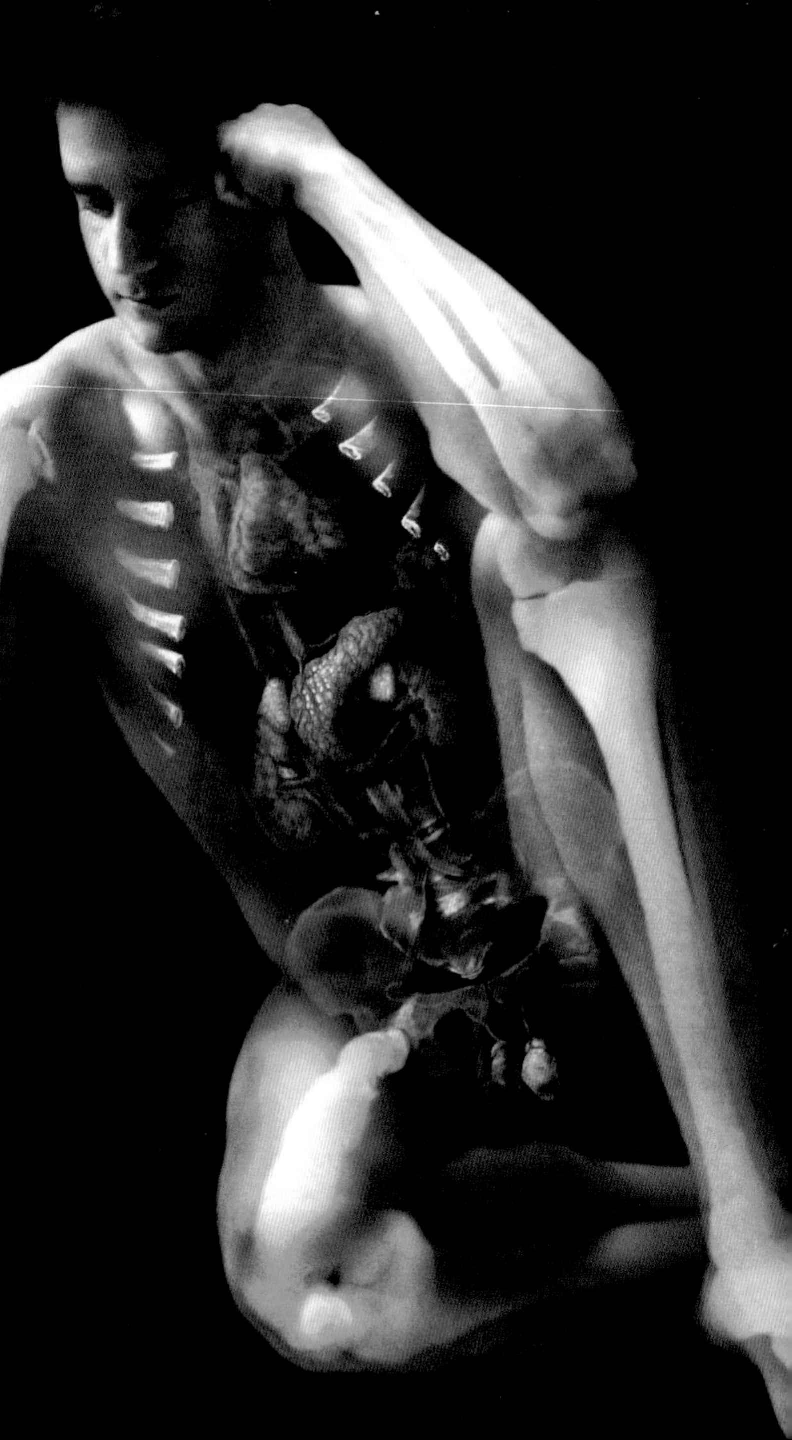

EXPANDING OUR HORIZONS

The future of treatments for sexual dysfunction is potentially as varied as the sex act itself. As researchers gain new insights into how sex works, they are developing treatments that affect sexual performance at all levels. Testosterone replacement therapy, in new and unique forms, not only may be used to enhance the sexual desire of men and women, but also may someday be used as a form of male contraception. Novel psychoactive drugs, or gene therapies, may be developed to directly alter the pathways of arousal and response. Innovative suppositories and topical creams could encourage erection in men,

A VAGINAL RING MIGHT DELIVER MICROBICIDES TO TARGET STDS.

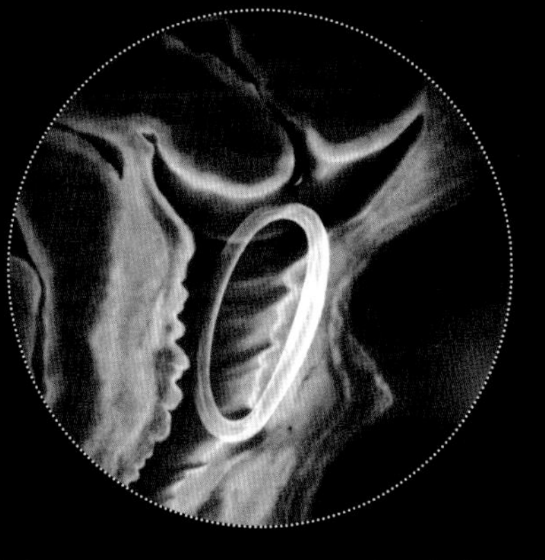

NEW POLYMERS COULD LEAD TO THE CREATION OF A CONDOM IN LIQUID FORM.

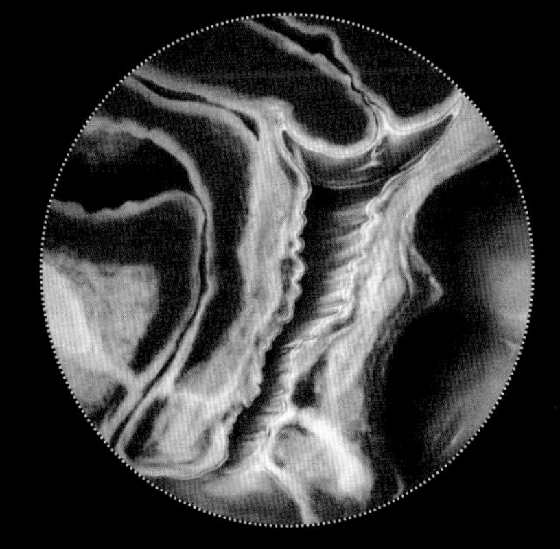

increase lubrication in women, and redefine the role of lubricants in sex.

Sex in the future will be safer as well. Microbicidal creams are being developed that kill not only sperm, but STD pathogens on contact. Other scientists are inventing ways to produce and deliver anti-STD antibodies directly into the vagina. These little bullets may finally provide women with a way to protect against disease without also preventing pregnancy. With a vaccine in the pipeline to prevent cervical cancer, and many of these products in clinical trials, it seems like tomorrow is already today.

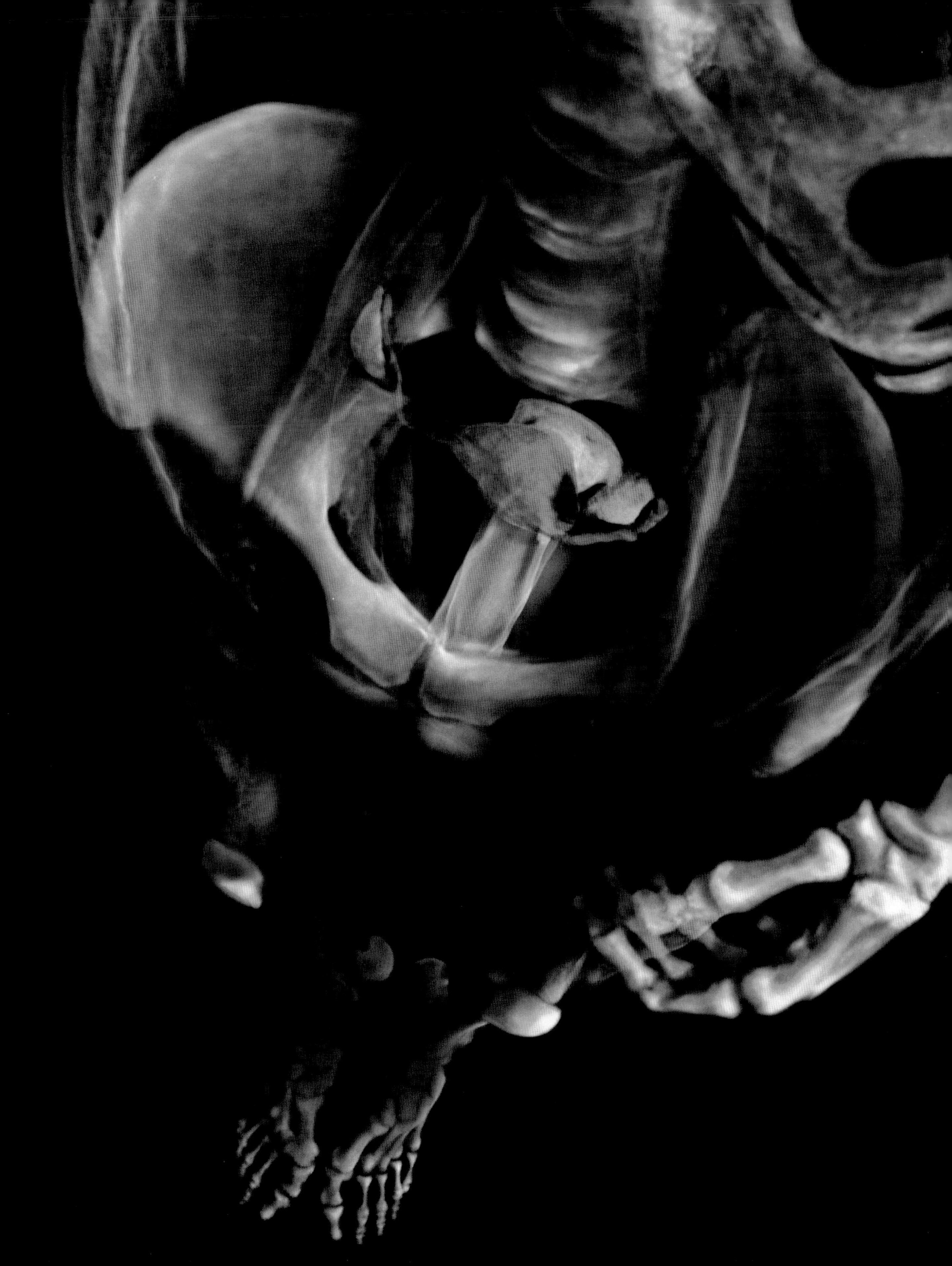

Against Diseases here,
the strongest Fence,
Is the defensive Virtue,
Abstinence.
—BENJAMIN FRANKLIN

6

LIVE, LAUGH, LOVE: LIFESTYLE SOLUTIONS

Everyone has a right to a healthy and happy sex life. It's important to remember, however, that how you live your life affects your sex life. So love safe, live well, and have fun.

LET'S HAVE A CONVERSATION

MOVE IT OR LOSE IT: The more you hate to get up in the morning, the harder it is to *get up* in the morning. If you want to be turned on, don't turn over. Get out of bed, get moving, and get your blood pumping. . . everywhere.

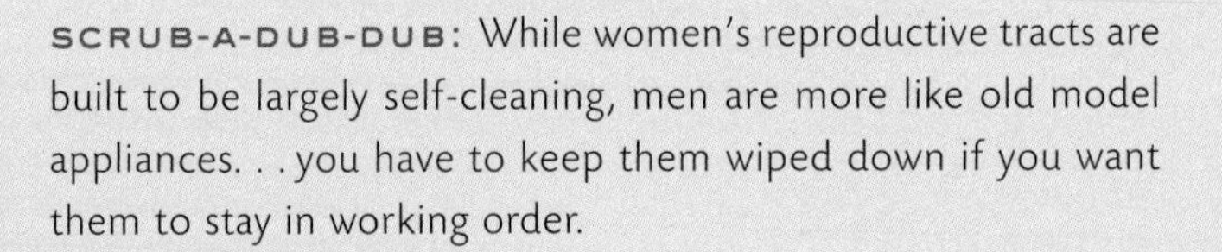

SCRUB-A-DUB-DUB: While women's reproductive tracts are built to be largely self-cleaning, men are more like old model appliances. . . you have to keep them wiped down if you want them to stay in working order.

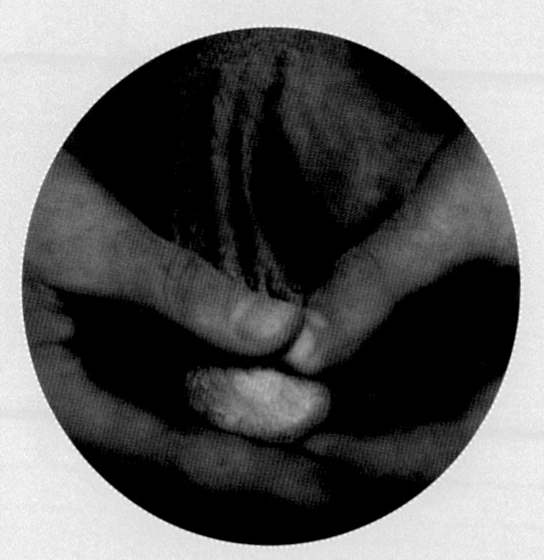

IN YOUR OWN HANDS: Ever since we were children, we've been told that an ounce of prevention is worth a pound of cure. It's certainly true when it comes to health—the sooner you discover a problem, the simpler it is to make it go away.

ABSTINENCE IS AN OPTION: Having an active sex life is a choice, and some people may choose to put their energy elsewhere.

A HEALTHY CARDIOVASCULAR SYSTEM IS ESSENTIAL TO SEXUAL FUNCTION IN MEN AND WOMEN. ANYTHING THAT CAN BE DONE TO MAINTAIN IT WILL IMPROVE SEXUAL HEALTH.

ON THE RIGHT TRACK

Physical health and sexual health are tightly linked. Stress, nutrition, and exercise all affect sexual performance and function. Because a healthy cardiovascular system is essential to sexual function in men and women, anything that can be done to maintain it will also improve sexual health. Eating healthfully, reducing your stress levels, and exercising regularly can restore sexual function in people whose sexual function has been impaired by excess weight or poor overall health.

Stamina isn't just for runners. Sex is a good form of exercise: doing it steadily for an hour would burn about 180 calories in an average-sized individual. Still, it's not nearly as efficient as walking (240 calories/hr), running (600-1200 calories/hr), or even taking your partner swing dancing (400 calories/hr). So if you want to get in shape for great sex, it's probably worth participating in some activity outside the bedroom.

CLEANLINESS IS A VIRTUE

The foreskin may be a tiny piece of skin, but it engenders an amount of controversy enormously disproportionate to its size. Circumcision—removal of the foreskin—is a practice common to many religions. The reasons for its recommendation range from hygienic to ritualistic, and in some degree it is a practice common to many if not most of the religions of the world.

Circumcision is also a consequence of the medicalization of pregnancy and birth. In the United States, where hospital births are almost universal, more than 50 percent of male infants are circumcised. However, in recent years, there has been a movement that decries infant circumcision as a cruel mutilation that diminishes sexual capacity and has no real benefit. Some insurance companies have stopped paying for the procedure, and the medical societies of many countries have criticized the practice as unnecessary surgery for routine cases. Today, some men are even undergoing procedures to stretch the skin of the penis to create a new foreskin.

Uncircumcised men are at increased risk of certain sexually transmitted diseases compared to circumcised men. This is mostly because the foreskin provides additional surface area for infection, and it also creates a warm, moist environment in which bacteria can grow. For this reason, some doctors recommend that parents gently retract their baby boy's foreskin and cleanse the head of the penis once a week with water, in order to reduce the risk of urinary tract infections, and that they teach their son to do this for himself when he grows older. Although circumcision is normally an elective procedure chosen by the parents of an infant, it can be necessary for some men who experience foreskin pain, inflammation, or other problems.

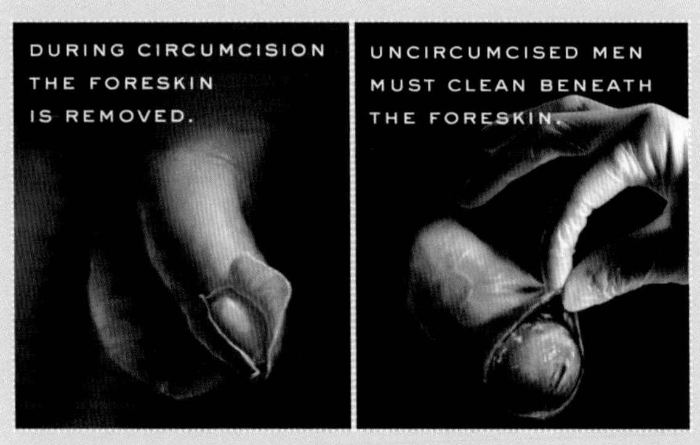
DURING CIRCUMCISION THE FORESKIN IS REMOVED.
UNCIRCUMCISED MEN MUST CLEAN BENEATH THE FORESKIN.

"Jock itch" is a fungal infection of the genitalia, and is also called *tinea cruris* or ringworm of the groin. It almost always occurs in adult men, thrives in warm wet places, and can be worsened by friction from clothes. It's also contagious—skin-to-skin contact can transmit it, as can contact with unwashed clothing—or even towels being snapped around in the locker room. Mostly it causes intense itching of the upper thighs, but it can also spread to the anus and cause itching there as well. Treatment is both topical—applied to the skin—and behavioral, since the skin needs to be kept warm and dry. Unfortunately, because it requires a change in lifestyle, jock itch frequently takes much longer than other ringworm infections to heal.

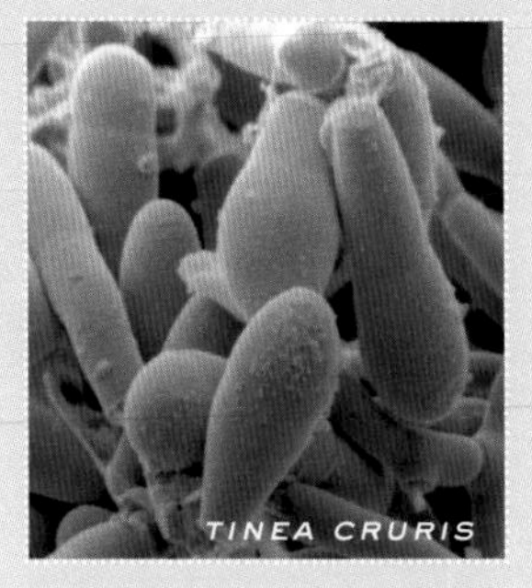
TINEA CRURIS

Urinary tract infections (UTIs) are a common irritation for both men and women. When people are healthy, urine is a sterile, bacteria-free fluid. However, growth of bacteria in the bladder, urethra, colon, or other parts of the gastrointestinal tract can cause this to change. UTIs occur more frequently in women because the shorter length of their urethra makes it easier for the bacteria to traverse, but the symptoms are largely the same for both sexes.

When you have a UTI, you constantly feel you need to urinate. When you do, not much urine will come out, and yet you'll still feel like you have to urinate. You may also feel a burning sensation when urinating. More serious infections can lead to fever, nausea, and even back pain if the bacteria ascend to the kidneys.

Treatment usually requires going to the doctor for antibiotics, but prevention is, as always, the simpler route. Drinking plenty of water, urinating before and after sex, and using a condom can prevent transmission of bacteria back and forth between you and your partner. This will help to prevent UTIs from developing in the first place—as will drinking cranberry juice, at least for women. It may be because of natural antibiotic properties of the juice, or it may simply be its acidity, but studies have demonstrated that drinking cranberry juice actually reduces the number of urinary tract infections experienced by women. And a tasty, healthy drink beats a doctor's visit any day.

For some women, there is no health condition that causes as much cursing and grumbling as a yeast infection. Although largely harmless in the grand scheme of things, yeast infections are incredibly annoying, causing not just discharge, but intense itching and also pain during sex. Yeast infections are easily curable, but just like some relatives, they tend to visit too often: women who are prone to yeast infections tend to get them regularly—sometimes as frequently as once a month, when their hormones make their vagina ripe for a takeover. As hormone production changes, the vaginal walls thicken and there are more nutrients in their cells to nourish the yeasts. During the first seven days of the menstrual cycle, the vaginal pH is at its highest and antibodies are at their lowest, leaving women particularly susceptible to infection. There is a reason why drugstores offer multi-packs of the various over-the-counter creams and pills formulated to combat these infections.

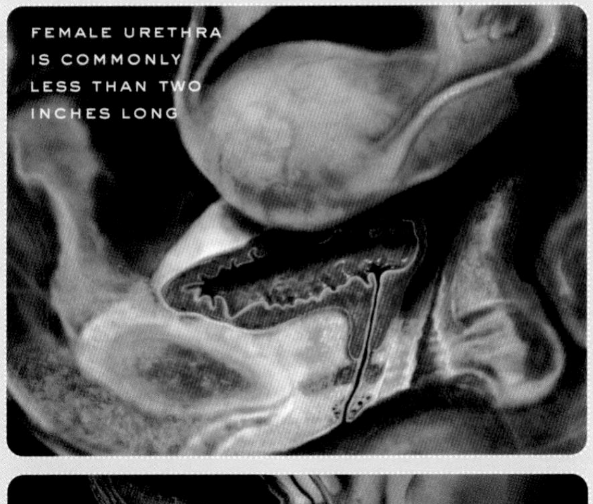

FEMALE URETHRA IS COMMONLY LESS THAN TWO INCHES LONG

MALE URETHRA IS COMMONLY NINE INCHES LONG.

No one really knows why some women never get a yeast infection, while it seems that others have helped to put the children of anti-fungal manufacturers through college, but it's important that women who have repeat episodes make certain that they are correctly diagnosed. Other conditions, such as lichen sclerosis, can masquerade as yeast infections, and frequent yeast infections could be a sign of a more serious infection, such as HIV. Although prevention of recurrent infection is in many ways a matter of luck, it's always safe to go back to the basics. Make certain that your vulva is rinsed clean of any soap and patted dry after you take a shower, stick to the cotton instead of the fancy underpants, and wipe from front to back. Many women with frequent yeast infections eat live culture yogurt to try and replenish the healthy vaginal bacteria that could guard against the disease, and this folk cure is soundly based in theory. Some women have also had success with reducing the amount of refined sugars, carbohydrates, and fermented products in their diet. Reducing refined sugar consumption takes away the yeasts' food; reducing carbohydrates and fermented products cuts down on their food source.

YEAST CELLS GROWING ON THE SKIN

BREAST SELF-EXAM IS AN IMPORTANT TOOL FOR THE EARLY DETECTION OF CANCER

CERVIX WITH CANCEROUS CELLS

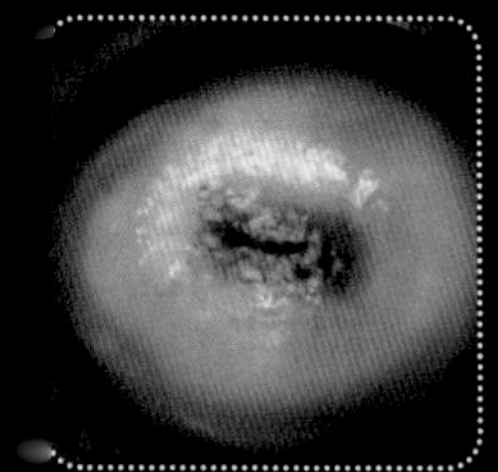

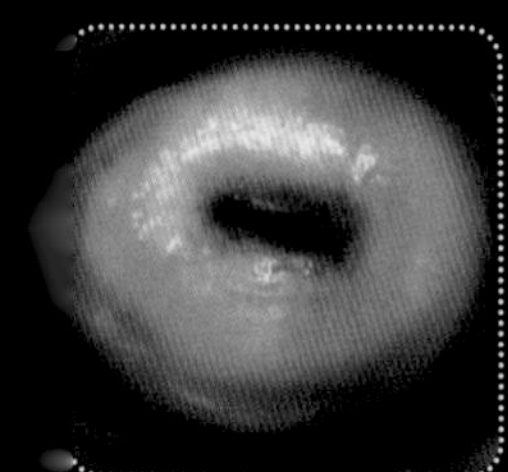

HEALTHY CERVIX

DURING A PAP SMEAR EXAM, A SWAB IS USED TO TAKE SAMPLES OF CERVICAL CELLS FOR ANALYSIS.

SELF AWARENESS

A woman's best source for information about her health is to listen to her own body. Paying attention to physical changes that are occurring within you can alert you to the early signs of serious illness. If you're bleeding between periods, your doctor might schedule an ultrasound to try to view the problem. If you have intense abdominal pain, she might need to schedule surgery to find out what's wrong.

Knowing what's normal for you can be useful in alerting you to what you need to bring to your doctor's attention, but there are some basic tests that every woman should have done—either by her doctor or herself.

Breast self-exams are an important tool for detecting changes in the breast that may signal the presence of cancer. You know your body best, and so you may be the first to notice if something changes or goes wrong. Doctors may be better at finding lumps in your breast tissue during your regular appointment, but you live with your breasts every day, and can check them accordingly. Some women perform a breast exam once a month when they're in the shower—it's as normal for them as washing their hair. Although not all breast lumps are malignant, they all should be evaluated by a doctor. Once a woman reaches the age of 40, her physician will probably recommend that she get an annual mammogram, in case normal age-related changes in her breast tissue are concealing something that she can't afford to miss. If a woman has a high-risk of breast cancer because she carries the breast cancer genes or has a family history of breast cancer, her physician may recommend annual mammograms at an earlier age.

Although HPV infection is the cause of most, if not all, cases of cervical cancer, screening for HPV is

percentage of women infected with the virus will eve develop cancer, and that could happen days or decade after initial exposure. That is why Pap smears are one o the most important preventative exams in women' health. During a Pap smear, the doctor takes a sample of the cells from a woman's cervix to check for change that could signal the onset of cervical cancer.

Every woman who has ever been sexually active even if it was 30 years ago, should get a Pap smear a least once every three years, because the virus tha causes cervical cancer can lie dormant for decades For women who have ever had cell changes on thei cervix, or who are in other high risk groups, Pap smears are recommended as a yearly activity. They're not that expensive, not that time-consuming, and they can save your life. No woman should ever die of cervical cancer, and those who are regularly screened don't.

The problem with Pap smears is that although the sample is easy to acquire, the actual testing requires practice and expertise. In some countries there is neither the money nor the technical expertise to provide every woman with regular Pap smears and in these countries cervical cancer is a leading cause of death. Fortunately, new, simple, and inexpensive tests for cervical cancer are being developed that can be performed by any trained member of the community. The most interesting one is based on the fact that when you apply simple vinegar, o acetic acid, to the cervix, healthy cells stay pink and unhealthy cells turn white. This can be seen easily through the magnifying lens known as a colposcope and thousands of women can be screened quickly and inexpensively. Only the women with potentially dangerous cervical changes are then referred to a

Smart men go to the doctor—and not just when they have problems with stress, their heart, or their head, but for regular checkups that include their penis, testicles, and prostate. The regular STD screenings, for example, that women get at their yearly gynecologist visit should also be received by men—even if they have to ask for them. Fortunately, these days, men don't need to suffer the discomfort and embarrassment of having a swab inserted in their penis to test for most STDs. Chlamydia and gonorrhea can now both be tested by using a urine test that looks for bacterial DNA, and syphilis and HIV are both diagnosed by blood tests.

MONITORING YOUR HEALTH

Men also need to take themselves in hand. . .at least once a month. Testicular self-exam is the best method for early detection of testicular cancer, and it can also help men discover other benign growths that may or may not benefit from treatment. From approximately the age of 15, men need to regularly examine their testicles for any change in size, pain on manipulation, or the presence of abnormal lumps. Knowing what to feel for requires practice, because there's a lot of stuff down there. If you can recognize what structures should be there, such as the epididymis, you will also recognize what shouldn't, and early detection of testicular cancer

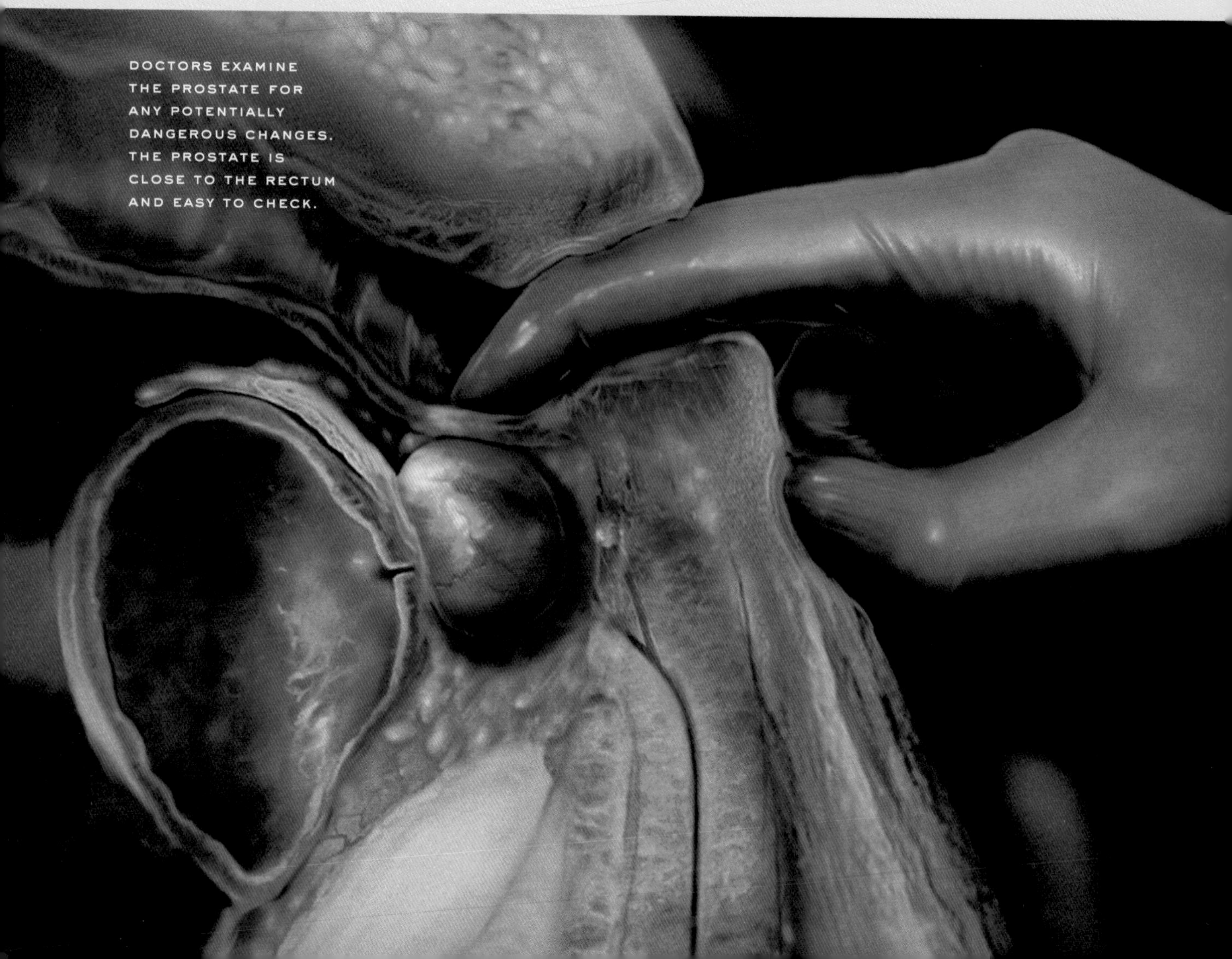

DOCTORS EXAMINE THE PROSTATE FOR ANY POTENTIALLY DANGEROUS CHANGES. THE PROSTATE IS CLOSE TO THE RECTUM AND EASY TO CHECK.

provides the best chance of cure.

OK, you heard it here first: the digital rectal exam has nothing to do with the Internet. Instead, it's a far more personal procedure in which a doctor inserts his lubricated finger a few inches into your anus to see if there's anything wrong with your prostate, anus, or rectum. This is necessary for the doctor to detect early signs of prostate enlargement as well as hernias and other bowel problems.

The PSA blood test is a relatively recent innovation that allows doctors to chemically screen for changes in the size of the prostate. PSA, or prostate specific antigen, is a protein made by the cells of the prostate. PSA increases in concentration throughout the body as the organ increases in size. The blood test does not, unfortunately, make the rectal exam obsolete, since many men experience benign growth of the prostate as they age, and the doctor still needs to check manually for pathogenic changes. It is also possible for a perfectly healthy person to have abnormally high PSA levels, or for individuals with prostate cancer to see no change in PSA levels. A PSA test and digital exam, as screening tests, are recommended yearly for all men over 50, and they should be started at 45 for African American men and other individuals at increased risk for prostate cancer.

Self-love, my liege, is not so vile a sin as self-neglecting.
—WILLIAM SHAKESPEARE, *Henry V*

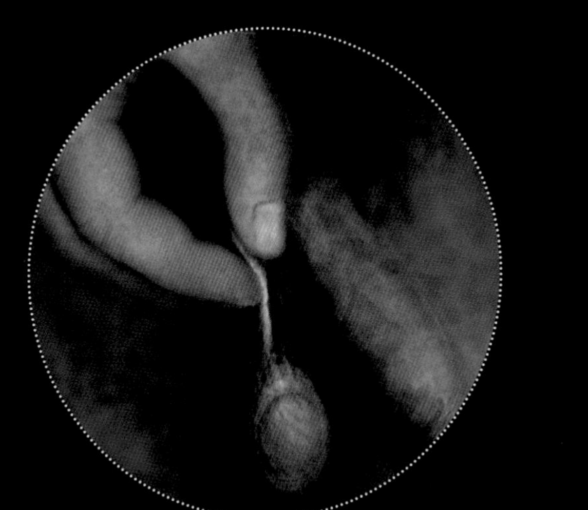

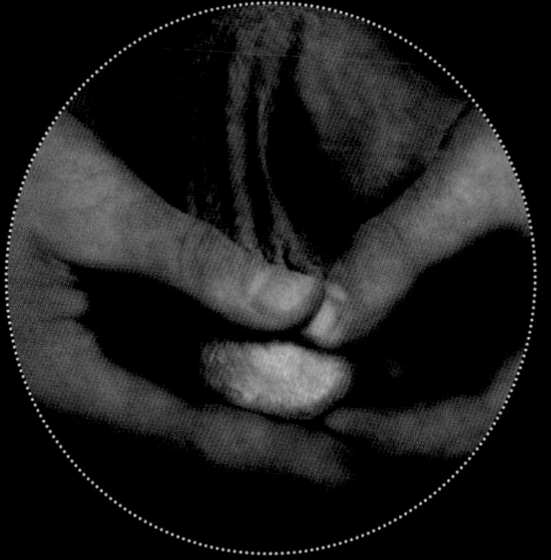

AT LEAST ONCE A MONTH IT IS IMPORTANT TO CHECK THE SPERMATIC CORD AND THE TESTICLES FOR ANY LUMPS OR CHANGES THAT COULD INDICATE THE PRESENCE OF DISEASE.

SEX: YOUR CHOICE

Abstinence—choosing not to engage in sexual contact with another person—can be a very powerful personal choice. Some people see it as the only moral option for unmarried men and women; others see it as a choice that can free them temporarily from physical distractions or preoccupations in order to pursue other avenues of personal growth.

People choose abstinence for many reasons, and at many different times of their lives. They may choose it because of religious or moral teachings that prohibit sexual contact outside of a strictly limited set of circumstances. They may choose it because they're more worried about STDs and pregnancy than interested in sexual expression. They may choose it to improve their semen quality in the hope of achieving pregnancy with their partner. All reasons are equally valid, and no one should be judged for choosing to be abstinent—just as they shouldn't be judged for choosing to have responsible sex.

Individuals have many different definitions of abstinence. For some people, being abstinent means no erotic contact more extreme than holding hands or kissing. For others, it means everything except vaginal intercourse, up to and including oral and anal sex. No health professional, parent, or teacher would consider the second definition to be abstinence, but knowing that some people consider it to be demonstrates the importance of clear and specific communication.

DECISIONS, DECISIONS

Oral sex is sex. Let's say it again: oral sex is sex. And while it can't get you pregnant, it can spread most of the same diseases that any other type of sex can.

In the U.S., there has been an upswing in the number of teenagers having oral sex, particularly casual oral sex. One reason why teenagers are doing it is that they think that oral sex is safe. They're wrong. The biology of your mouth is very similar to the biology of your genitals: both are warm, wet spaces lined with mucous membranes, and almost any disease you can catch "down there" you can also catch through oral exposure. HIV, herpes, HPV, gonorrhea, syphilis, and even yeast infections can all be acquired through the mouth. Since you can't tell if a person is infected simply by looking at them, smelling them, or even tasting them, it's important to always use a latex barrier when engaging in oral sex—either a condom when performing fellatio on a man, or a flattened condom or dental dam when performing cunnilingus on a woman.

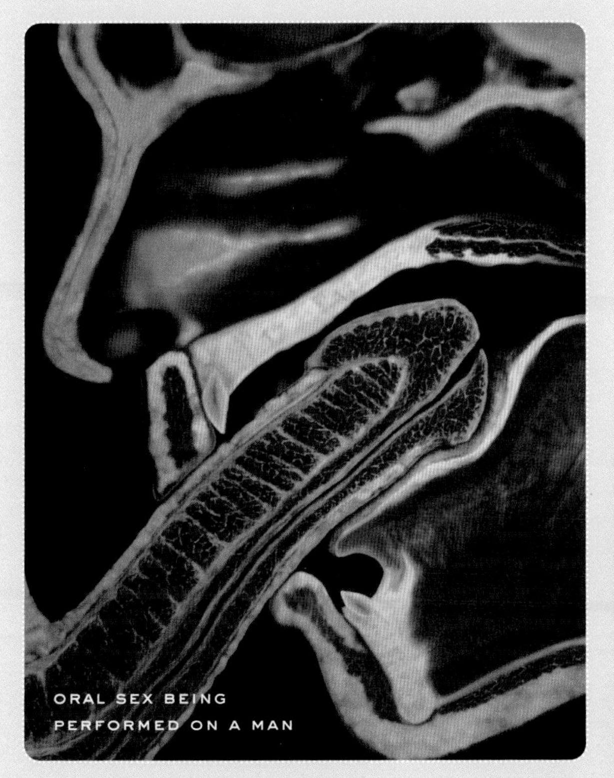

ORAL SEX BEING PERFORMED ON A MAN

It doesn't matter what acts you define as sex, and what acts you define as friendly, the important thing is that your sex life should be safe as well as fun. And if your partner is unwilling to practice safe sex with you, you should reconsider your choice of partners. Safe fun isn't an oxymoron . . . sometimes safe makes things even better, not just by eliminating worry but by increasing your sexual repertoire. Sex toys can be an exciting component of safer sex, just put a condom on them or disinfect them thoroughly before sharing. Buy flavored condoms or lubricants for oral sex, but make sure they're labeled for actual use and not just novelties (novelty varieties are not manufactured for safe sex purposes), to make protection a tasty after-dinner treat. Experiment with activities that don't involve exchange of potentially infectious fluids; they can be just as satisfying as those that do. Sex can be a wonderful experience, and it only gets better with a little advance planning.

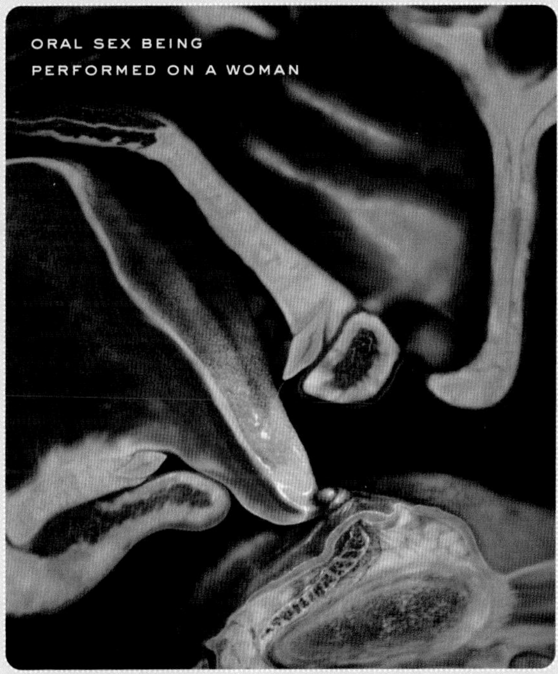

ORAL SEX BEING PERFORMED ON A WOMAN

**I've tried several varieties of sex.
The conventional position
makes me claustrophobic and
the others give me a stiff neck or lockjaw.**
—TALLULAH BANKHEAD

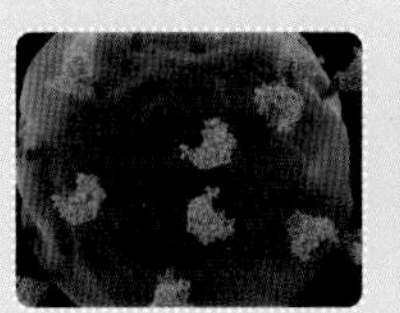

HIV

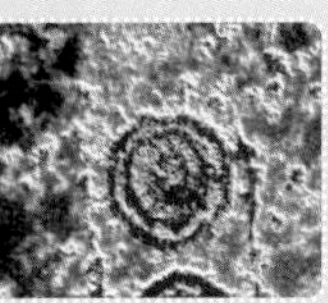

HERPES

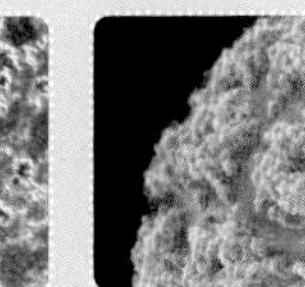

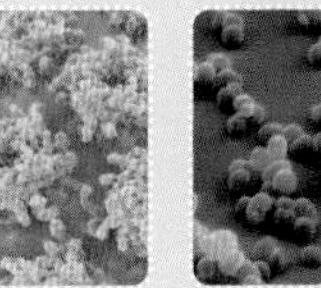

HPV

GONORRHEA

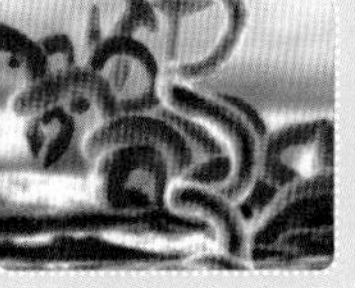

SYPHILIS

YEAST INFECTIONS

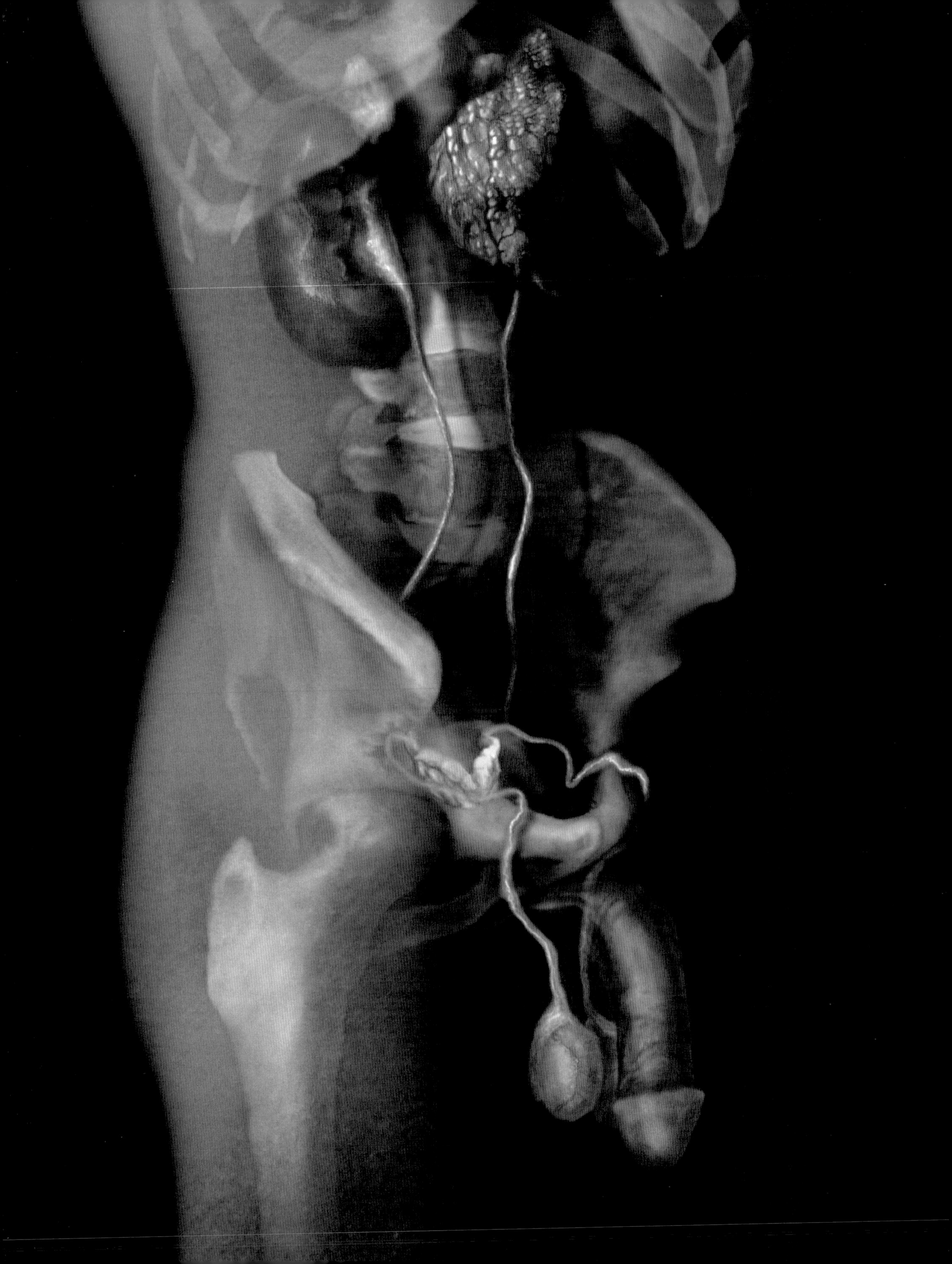

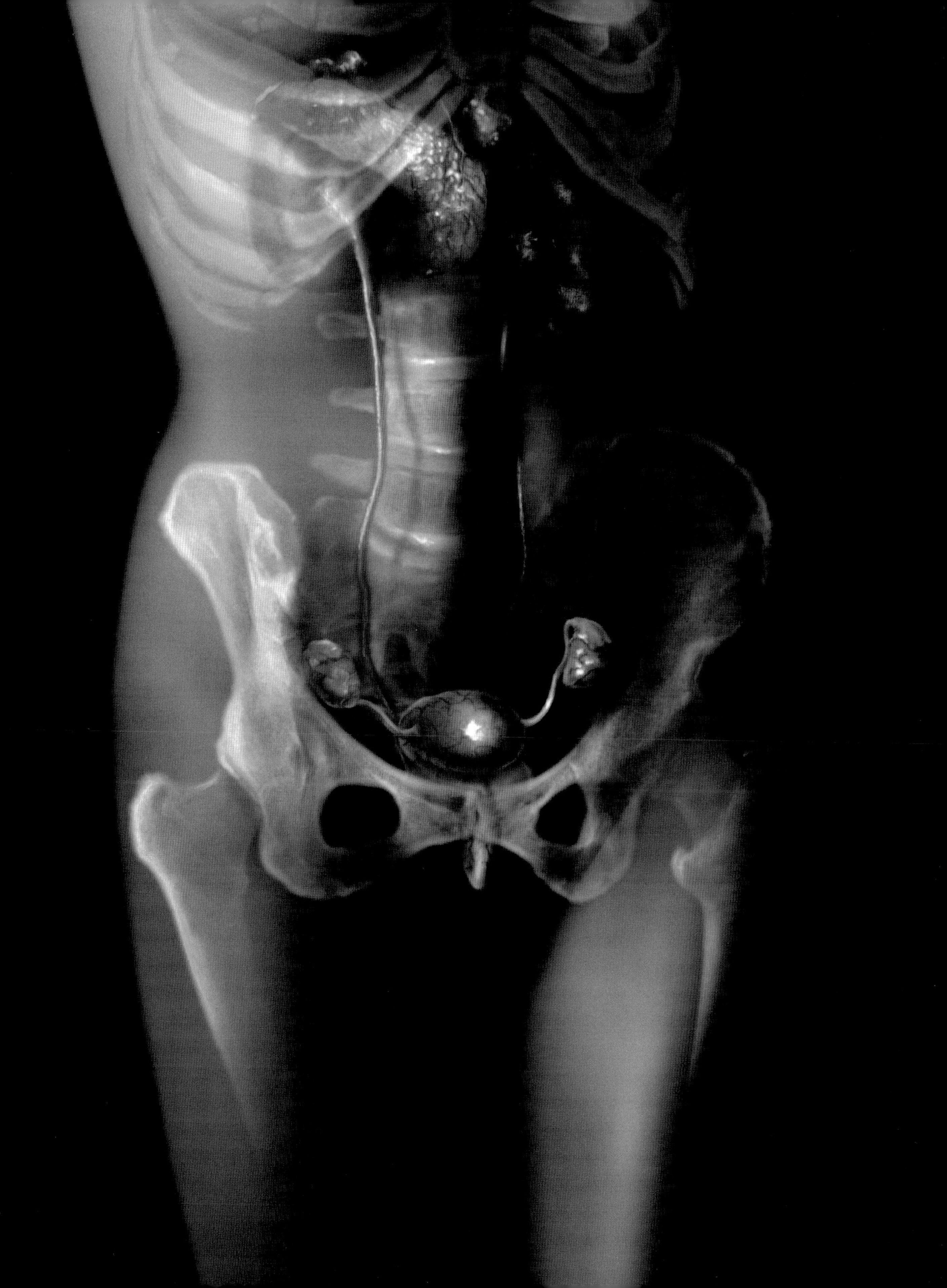

FOOTNOTES

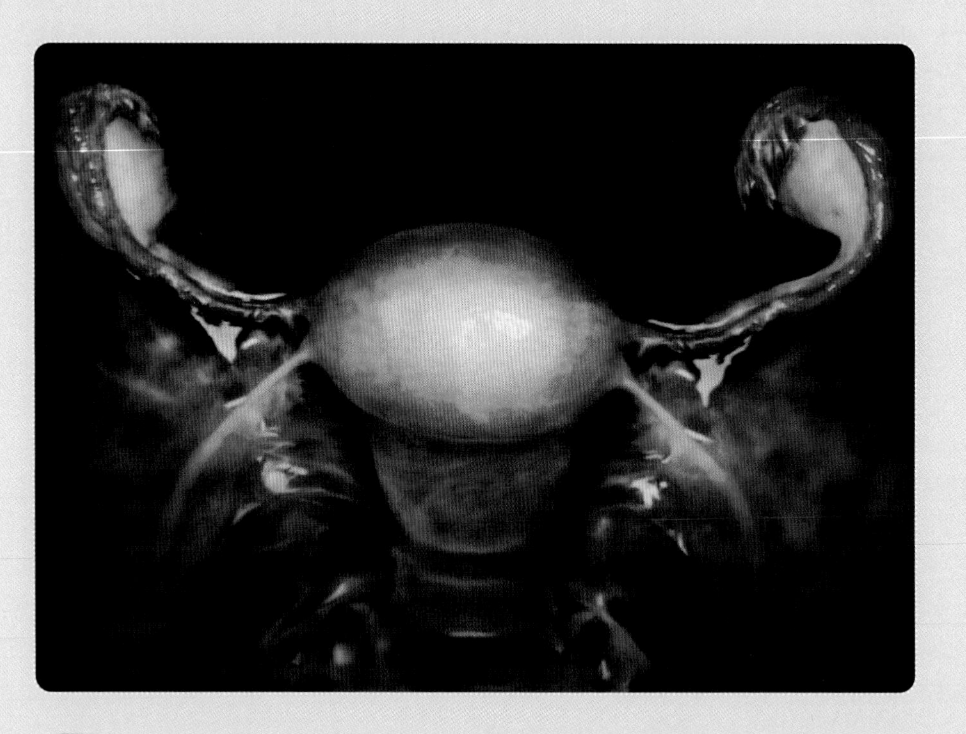

[1] Ward, P. *Is it a boy or a girl?* Great Falls, Va.: Discovery Channel. Cable broadcast. March 26, 2000.

[2] Manning, J. T. 2002. *Digit ratio: A pointer to fertility, behavior, and health*. New Brunswick, NJ: Rutgers University Press.

[3] Jankowiak, W., and E. Fisher. 1992. A cross-cultural perspective on romantic love. *Ethology* 31:149–155. Updated in Jankowiak, W. 1995. Introduction. In *Romantic Passion: A Universal Experience?* ed. W. Jankowiak, 1–20. New York: Columbia University Press.

[4] Tuiten, A., J. Van Honk, H. Koppeschaar, C. Bernaards, and J. Thijssen. Time course effects of testosterone administration on sexual arousal in women. *Arch. Gen. Psychiatry* 57:149–153.

[5] Guay, A. T. 2001. Decreased testosterone in regularly menstruating women with decreased libido: a clinical observation. *J. Sex Marital Ther.* 27:513–519.

[6] Clark, A. P. 2004. Self-perceived attractiveness and masculinization predict women's sociosexuality. *Evol. Hum. Behav.*, 25:113–124.

[7] Langley L. L., ed. 1973. *Contraception*. Stroudsburg, Pa.: Dowden, Hutchinson and Ross.

[8] "Sex can be a big yawn for patients who take anti-depressant drug." *The Daily Telegraph*, 4 September 1995.

GLOSSARY/INDEX

A

abstinence The practice of choosing not to partake of an item or an activity. In this book, used to refer to choosing not to engage in sex. (130, 133, 142)

adrenal cortex The steroid hormone-producing portion of the adrenal glands that are located on top of the human kidney. (22, 43)

adolescence The period of life that begins at puberty and ends at physical maturity. (15)

Adam's apple A lump seen in men's necks. It consists of the front edge of the thyroid cartilage of the larynx. (18)

anti-mullerian hormone (AMH) A hormone secreted in the male fetus that prevents the development of the upper vagina, cervix, uterus, and fallopian tubes. (10)

androgen The steroid hormones responsible for the development of male characteristics. (12, 13, 43)

antidepressants Several classes of drugs used to treat depression. (58, 98, 112)

arousal A state where the activity of a physical system increases. Specifically, in this book, it refers to sexual arousal, which involves the excitement of several systems. (5, 18, 28, 30-32, 40, 42, 44, 46, 47, 49, 56, 58, 63, 64, 67, 69, 71, 74, 77, 98, 113, 122, 124, 126)

autonomic nervous system The part of the nervous system responsible for the control of involuntary, or unconscious, actions. It is divided into the sympathetic and parasympathetic nervous systems. (62-63)

B

biopsy The removal of a tissue specimen from an organ to test for disease. (101, 119)

breast The milk-secreting glands of the female body and their surrounding tissue. In males, the tissue surrounding the rudimentary glands. (14, 15, 17, 19, 40, 82, 83, 94, 117, 139)

blood The fluid that brings nutrients to tissues and removes waste products via the circulatory system. It consists of plasma and several types of cells. (5, 8, 14, 17, 21, 31, 38, 43, 44, 55, 56, 58, 63, 65, 67, 69, 71, 72, 74, 76, 77, 90, 97, 100, 101, 112, 113, 118, 119, 122, 124, 125, 132, 140, 141)

blood vessels The arteries, veins, and capillaries that provide a pathway for blood to circulate throughout the body. (21, 38, 43, 71, 72, 77, 101, 113, 118)

buffer Any solution that can maintain its pH (proton concentration or level of acidity) effectively when other acids or bases are added. (34, 38)

C

cancer A type of disease where cells in the body divide uncontrollably or abnormally, and can invade other tissues through a process called metastasis. (14, 77, 82, 83, 89, 91-97, 101, 104, 106, 116, 118-120, 127, 138-141)

central nervous system (CNS) The brain and spinal cord. (56, 67)

cervix In this book, the donut-shaped area that separates the uterus from the vagina. Also means "neck." (16, 31, 36, 63, 85, 92, 93, 111, 138, 139)

conception The formation of a zygote when the egg and sperm successfully combine. (5, 7)

condom A device designed to cover the penis and contain semen upon ejaculation. The female condom performs the same function, but is inserted into the vagina instead. Both types function to protect against conception and many sexually transmitted diseases. (65, 84, 89, 91, 104, 108, 109, 110, 112, 127, 137, 142, 143)

contraception A method or device designed to prevent conception from occurring during vaginal intercourse. (108, 109, 111, 126)

chromosome A structure made of DNA that contains genes. Humans have 23 pairs of chromosomes. (7, 8, 10)

cholesterol A compound produced by the liver, and found in various foods, that is important for the synthesis of steroid hormones, as well as the maintenance of certain cellular structures. In excessive amounts it can raise the risk of disease. (21, 77)

clitoris A highly innervated erectile organ found at the top of the vulva in women. It is homologous to the penis in the male. (12, 13, 18, 30-32, 36, 40, 57, 63, 71, 74, 124)

corpora cavernosa Two chambers of spongy tissue that run the length of the penis. Erection occurs when they fill with blood, expanding and becoming stiff. (31, 41)

corpus spongiosum A single column of erectile tissue in the penis that surrounds the urethra. It becomes filled with blood during erection. (31, 41)

cortisol The "stress hormone." Levels of cortisol increase during times of stress. (58)

Cowper's glands Two small glands behind the prostate that contribute to the production of semen. (32, 38, 39)

cytoplasm The complex liquid that is found outside the nucleus of the cell. (8)

cyst A sac found inside the body which may be filled with liquids, semi-solid material, or gas. A cyst may or may not be abnormal. (99-101, 107, 118, 119)

D

diabetes Several metabolic abnormalities in which the body does not correctly process glucose (sugar). Diabetes is so named because of the excessive urination by which it is characterized. It can lead to systemic complications if not properly controlled. (55, 72, 74, 75, 77)

dihydrotestosterone (DHT) A highly active derivative of testosterone, produced by 5-alpha reductase. (12, 13, 19, 22)

deoxyribonucleic acid (DNA) The cellular structure that encodes hereditary characteristics (genes). (8, 92, 140)

E

egg The female gamete. It contains half the genetic material necessary to create a new organism. (4, 7, 8, 16, 39, 111, 126)

endothelial dysfunction A type of damage to the cells lining blood vessels, seen most commonly in individuals with high cholesterol, hypertension, or diabetes. It is one of the earliest measurable signs of cardiovascular disease. (77)

erection The stiffening of the penis during sexual arousal. (5, 18, 41, 60, 63, 64, 67, 69, 71, 72, 77, 100, 112, 113, 124-126)

erectile tissue A specific type of tissue that has evolved to become rigid when filled with blood. (30, 31, 70, 74, 100, 102)
erotic Having to do with sex and/or desire. (28, 44, 47, 65, 69, 142)
erogenous zones Parts of the body that are likely to cause sexual feelings when touched. These areas generally contain high concentrations of nerves, but vary with each individual. (29, 49)
embryo An organism in the earliest stages of development– before it has reached a structurally recognizable form. (2, 7, 8)
epididymis The tubes that carry sperm to the vas deferens. (11, 39, 101, 140)
erectile dysfunction A condition where men experience difficulty forming or maintaining an erection. (67, 71, 72, 77, 112, 124, 125)
estrogen A type of hormone, primarily produced in the ovaries, that is responsible for, among other things, the development of female secondary sexual characteristics and the growth of the long bones of the human body. (14-16, 18, 22, 43, 61, 65, 120, 122)
ejaculation The forceful expulsion of semen from the penis that usually occurs during orgasm. (14, 21, 30, 34, 36, 38, 39, 41, 43, 58, 63, 72, 106, 112, 113, 116, 117)

F

fetus The unborn offspring once it has developed to a stage where it structurally resembles the adult. In humans, this occurs at around eight weeks after conception. (8, 10-13)
fallopian tube The tubes through which the egg descends from the ovaries to the uterus. Fertilization usually occurs while the egg is in the fallopian tubes. (10, 61, 84, 108, 111, 117)
fertilization The joining of gametes to create a viable zygote. This creates a cell with a full complement of chromosomes, instead of the half complement seen in each of the gametes. (39)
foreskin The loose skin that covers the glans penis in uncircumcised men. (136)
functional magnetic resonance imagery (fMRI) A type of medical imaging that allows scientists and doctors to see what areas are activated in the brain by specific activities or stimuli. (32)
follicle stimulating hormone (FSH) A pituitary hormone that stimulates production of sperm in the testes, and growth of follicles in the ovaries. (14, 16, 18, 21)

G

gametes Cells with half the normal number of chromosomes, which can join together to produce a viable offspring. (21)
gene A section of DNA that produces a protein product. It is thought of as the primary unit of heredity. A chromosome contains many genes. (7, 8, 22, 94, 139)
genitals The reproductive organs, specifically the external sex organs (e.g. the penis, vagina, clitoris). (2, 13, 15, 43, 89, 142)
genital warts Refers to a disease where warts are present on the genitalia. Genital warts are caused by some strains of human papilloma virus. (89)
gonorrhea A sexually transmitted disease caused by the bacterium *Neisseria gonorrhoeae*. It is frequently asymptomatic, but can cause discharge from the cervix or penile urethra. (85, 86, 143)
glans The head of the penis or clitoris. (32, 36, 41, 112)
gonads The organs that produce gametes. Specifically the ovaries and testes. (10, 22)
genetics The study of heredity. (7, 8)
gonadotropin releasing hormone (GnRH) A hormone produced by the hypothalamus that causes the pituitary to release luteinizing hormone and follicle stiumlating hormone. (14)
guevedoches Literally "balls at 12." The term refers to a group of XY individuals with a genetic disorder that causes them to appear as female until puberty, at which point their appearance changes to male. (13)

H

herpes simplex (HSV) Either of two diseases caused by the herpes simplex virus that frequently have no symptoms but can cause skin blistering. These are known as oral herpes (HSV1/cold sores) and genital herpes (HSV2), although both viruses can infect either site. (88, 89, 91, 114, 142, 143)
hormone Any substance produced by the body that travels through the bloodstream and has activity somewhere else. (5, 8, 14, 17, 18, 21, 22, 43, 44, 49, 58, 60, 63, 69, 98, 107, 117, 121, 122, 127)
hymen A piece of tissue that partially or fully blocks entrance into the vagina and ranges in thickness. (14)
hypertension High blood pressure. (77, 113)
hysterectomy Surgical removal of the uterus. May include removal of the ovaries, although this is technically known as oophorectomy. (98, 117, 119)

I

immunoglobulins Also known as antibodies. These are proteins that bind with specific antigens. (38)
inguinal canal The passage through the abdominal wall in which the testes descend into the scrotum. It contains the spermatic cord in adult males, and the round ligament (which supports the uterus) in females. (11)
intercourse Sexual contact involving the genitalia of at least one party (e.g. anal intercourse, vaginal intercourse). (14, 30, 34, 36, 37, 67, 74, 108, 142)

L

labia Lips. In this book, specifically, the labia majora and minora – the tissue flaps surrounding the clitoris and vagina. (12, 13, 15, 30, 57, 63, 75)
larynx Also known as the voice box. Made of muscle and cartilage, it contains the vocal cords. (18, 19)
leydig cells The cells in the testicles that secrete testosterone. (21)
libido A word describing the extent of someone's interest in sex. (22, 43, 60, 72, 112)
limbic system A group of brain structures that are involved in the perception of and response to emotion. (44, 48, 58, 65, 69)
lubrication The addition of a slippery fluid to a process in order to reduce the effects of friction. In this book, specifically, it refers to the increase in vaginal secretions that occur to ease the friction of sexual intercourse. (31, 34, 38, 60, 61, 63, 71, 74, 77, 98, 109, 120, 124, 127)
luteinizing hormone (LH) A hormone produced in the anterior

pituitary that controls the production of testosterone in men, and stimulates ovulation and the creation of the corpus luteum in women. (16, 18, 20, 23)

M

macrophages A type of white blood cell responsible for the ingestion of bacteria and other substances. (21)
mammogram A screening test used to detect potentially cancerous changes in breast tissue. (94, 139)
masturbation Sexual stimulation of one's own body. Also can be used to refer to stimulation of a partner's body in the way that they would stimulate themselves (mutual masturbation). (14, 18, 83)
meiotic A type of cell division that produces 4 cells with half the parental number of chromosomes. In contrast, mitosis produces two cells, each with the same number of chromosomes as their parent cell. (21)
membrane A thin layer of material enclosing a structure, or separating one or more structures. (142)
menarche A woman's first menstrual period. (14)
menopause The time at which a woman's body ceases to menstruate. (117, 120)
menstrual cycle The series of hormone-controlled changes that occur on approximately a monthly cycle that prepare a woman for the possibility of becoming pregnant. Included are the release of an egg, the thickening of the uterine lining, and the shedding of the thickened lining if a pregnancy does not occur. (16, 17, 18, 21, 48, 49, 98, 137)
menses The shedding of endometrial tissue and blood that occurs at the end of the menstrual cycle if a woman has not become pregnant. (16)
mitochondria The small bacteria-like organelles that provide energy to a cell. They contain their own DNA, and mitochondrial DNA only comes from the mother, not the father. (8)
mucin The protein that gives mucus its viscous qualities. (38)
Mullerian ducts The ducts that in a female fetus will become the upper third of the vagina, cervix, uterus, and fallopian tubes. (10)
multiple sclerosis A chronic disease where nerve fibers lose parts of their insulating coating, and no longer function properly. It can cause muscle weakness, loss of coordination, and other sensory effects. (77)
muscle A type of tissue responsible for movement of physical structures. (18, 19, 21, 22, 31, 34, 36, 40, 60, 63, 71, 72, 74, 98, 112, 122)

N

nerves Fibrous structures that transmit sensory stimuli and motor impulses throughout the body. They can be thought of as a physical information superhighway. (29, 30, 36, 42, 43, 54, 57, 63, 66, 67, 69, 72, 75, 88, 98)
nipples The small structures at the center of the female breast that provide an outlet for the milk ducts. The equivalent non-functional structures exist on the male breast. (15, 19, 21, 31)
nitric oxide A compound that is involved in blood vessel dilation and other physiological activity. Extremely important during sexual arousal. (68, 71, 77)
nocturnal emission Ejaculation while a man is asleep. Also known as a wet dream. (18, 32)

O

olfactory Relating to the sense of smell. (48)
ovaries The organs in the female body that produce eggs and the sex hormones estrogen and progesterone. (10, 11, 14, 22, 43, 61, 77)
orgasm A release that occurs at the peak of sexual excitement. It is normally accompanied by ejaculation in the male and vaginal contractions in the female. (28, 30, 32, 34, 36, 40, 41, 43, 58, 64, 67, 68, 72, 98, 112)
oxytocin A hormone released from the posterior lobe of the pituitary gland that plays a role in orgasm. It also contributes to the ejection of milk from the breast during nursing and contraction of the uterus during labor and delivery. (68, 69)

P

Pap smear A diagnostic test used to detect the presence of precancerous or cancerous cells on the cervix. (92, 138, 139)
parasympathetic nervous system The part of the nervous system that is not the central nervous system. It is responsible for "rest and digest" responses. (62)
peripheral nervous system (PNS) The part of the nervous system that is not the central nervous system. It consists of the autonomic nervous system and the somatic nervous system (which contains the sensory nerves and those nerves responsible for motor stimulation). (56, 63)
Peyronie's disease A disease where scar tissue forms around the erectile bodies of the penis leading to deformed or painful erections. (67)
puberty The time during which a child's body matures and acquires full reproductive function. (5, 8, 13-15, 18, 19, 21, 22)
penis An erectile structure involved in the delivery of sperm during intercourse. It also contains the urethra through which urine is disposed of from the bladder. (14, 15, 20-22, 32-34, 38, 43, 65, 69, 71, 72, 74, 100, 102, 114, 126, 127, 138, 142)
pornography Media created to encourage sexual arousal. (29, 38)
pituitary One of the primary glands of the endocrine system, located at the base of the brain. It has numerous regulatory functions. (16, 21, 22, 43, 68, 77)
pheromones Odorless compounds that stimulate various physical processes. They have been shown to be able to synchronize women's menstrual cycles, and also play a role in the biology of attraction. (48, 49)
physiology The study of how living organisms function. (30)
pregnancy The state of carrying a developing fetus. The time between conception and birth. (13, 14, 16, 22, 86, 98, 108, 110, 127, 136, 142)
premature ejaculation Ejaculation during the sexual act at a time earlier than the man, or his partner, wishes. (58, 106, 112)
progesterone A pro-gestational hormone, that helps to prepare for and maintain pregnancy. It is made in the corpus luteum (in the ovaries) and the placenta. (16)
prostate A gland that contributes to the production of semen. It

contains many nerves, and can be pleasurable to stimulate. (14, 21, 23, 30, 36, 38, 96, 97, 116, 140, 141)
pre-menstrual syndrome PMS PMS is a group of symptoms, one or more of which many women experience during the two weeks before their menstrual period. These symptoms include breast tenderness, upset stomach, and mood changes. The more disabling version of this condition is known as premenstrual dysphoric disorder.

R

reproduction The sexual or asexual process of generating offspring. (7, 16, 22, 43, 44)
resolution The ending of an abnormal physical process. In this book, it refers to the body's process of returning to its unaroused state. (30, 40, 41, 60)
retrograde ejaculation A harmless condition where semen moves backwards up the urethra into the bladder instead of being expelled from the penis during ejaculation. (72, 116)

S

semen The ejaculatory fluid containing sperm and secretions from the prostate and other glands. (34, 38, 39, 90, 142)
scrotum The sac that encloses the testes. (11, 12, 13, 18, 21, 30, 60, 63, 100, 101, 108)
seminiferous tubules The tubes inside the testes in which sperm are formed and begin to mature. (21)
sex The state of being genetically male or female. Any of a number of activities associated with sexual intercourse or reproduction. (3-5, 7, 8, 10, 12-14, 16-18, 22, 23, 27-32, 34, 36, 38. 40, 41, 43, 44, 46-49, 52-56, 58, 60, 61, 63-69, 74, 77, 80-82, 84, 87, 89-91, 94, 98, 104, 107, 111, 113, 122, 126, 127, 131, 133, 135, 136, 139, 142, 143)
sexually transmitted disease (STD) A disease where the primary mode of transmission is through sexual contact. Usually transmitted through semen, vaginal secretion, or skin-to-skin contact. (14, 81, 82, 84, 85, 107, 136)
sperm The male gamete. Contains half of the genetic material needed to form a zygote. (4, 7, 8, 18, 20, 21, 30, 31, 34, 38, 39, 61, 96, 100, 101, 107, 108, 111, 116, 126, 127)
spermatic cord The structure that suspends the testes from the body. Can shorten or lengthen to control their temperature. Consists of the vas deferens, blood vessels, and other tissue. (31, 100, 101, 141)
spermatogenesis The process of generating sperm. (21)
seminal vesicle Glands next to the male urethra that secrete seminal fluid. (10, 11, 36, 38, 39)
steroids Any hormone affecting the growth or development of sex organs. Also, any of several fat-soluble organic compounds having as a basis 17 carbon atoms in four rings. (58, 122)
sympathetic nervous system The part of the autonomic nervous system that is responsible for preparing the body to respond to stressful or emergency situations–"fight or flight." Works in opposition to the parasympathetic nervous system. (63)
syphilis A primarily sexually transmitted disease caused by the spirochete *Treponema pallidum*. If left untreated it can cause systemic infection and even death. (85-87, 109, 140, 142, 143)
systemic circulation The flow of blood throughout the human body. (77)

T

tampon An absorbent tube inserted into the vagina to collect menstrual blood. (14, 16, 17)
testosterone The steroid hormone responsible for much of male sexual development, sexual interest in women, and other developmental functions. (11-13, 18, 19-23, 30, 43, 47, 58, 60, 61, 68, 97, 116, 122, 126)
testes The male reproductive glands that produce spermatozoa as well as testosterone and other androgens. Singular is testis. (100, 101)
testicle One testis. See above. (10-13, 18, 20, 21, 30, 31, 43, 60, 63, 77, 96, 97, 116-118, 140, 141)

U

uterus The hollow organ where the zygote implants, matures, and develops. (7, 10, 16, 17, 30, 31, 36, 40, 60, 61, 95, 98, 111, 117, 119)
urethra The duct through which urine is discharged from the bladder. In males also carries semen. (12, 36, 38, 39, 63, 85, 86, 136, 137)

V

vagina The tube-like structure leading from the external female genitalia to the entrance of the uterus. (10, 12-17, 30-32, 34, 36, 38, 40, 43, 60, 61, 63, 65, 67, 71, 74, 84, 86, 95, 98, 108, 113, 120, 127, 137, 142)
vas deferens The main duct through which sperm is carried from the testicles. (11, 36, 39, 108)
vasocongestion A localized increase in blood flow. (31, 32, 60, 113)
virgin A person who has not engaged in some range of sexual contact. Specific definition varies strongly by culture, from lack of any sexual contact to never having experienced vaginal intercourse. (14)

W

wet dream Also called nocturnal emission. Ejaculation while a man is sleeping. (18)
wolffian ducts The structures that, in the fetus, will become the vas deferens in males, and the ureter in both males and females. (10, 11)
womb Another term for the uterus. The place where the zygote matures and develops. (7, 8, 13)

X

X chromosome One of the two sex chromosomes, the X chromosome is present in both men (XY) and women (XX). The sex chrosomes are the only chromosome pair in humans that don't contain the same set of genes. (7)

Y

Y chromosome One of the two sex chromosomes, the Y chromosome is only present in men (XY). If the Y chromosome doesn't function properly, XY individuals may appear to be female. (8, 10)